Linear Algebra

for use with

Written by MIROSLAV LOVRIĆ

NELSON EDUCATION

NELSON EDUCATION

Linear Algebra
by Miroslav Lovrić

for use with *Calculus for the Life Sciences*, Second Canadian Edition
by Frederick Adler and Miroslav Lovrić

Vice President, Editorial Higher Education:
Anne Williams

Publisher:
Paul Fam

Executive Editor:
Jackie Wood

Marketing Manager:
Leanne Newell

Developmental Editor:
Suzanne Simpson Millar

Technical Checker:
Caroline Purdy

Content Production Manager:
Claire Horsnell

Copy Editor:
Heather Sangster at Strong Finish

Design Director:
Ken Phipps

Managing Designer:
Franca Amore

Cover Design:
Martyn Schmoll

Cover Image:
iLexx/iStockphoto.come

Printed and bound in Canada
1 2 3 4 17 16 15 14

For more information contact Nelson Education Ltd., 1120 Birchmount Road, Toronto, Ontario, M1K 5G4. Or you can visit our Internet site at http://www.nelson.com

ISBN-13: 978-0-17-657137-5
ISBN-10: 0-17-657137-X

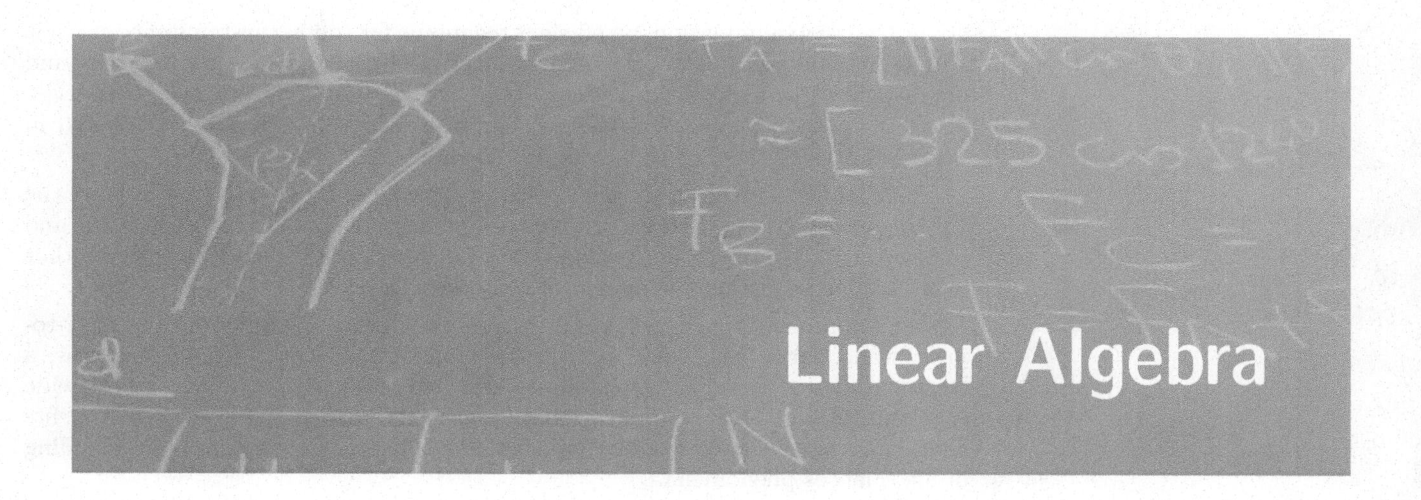

Linear Algebra

When we think about what mathematics we can use to model phenomena in the life sciences (and more broadly), linear algebra is not the first thing that comes to mind. Calculus, probability, statistics, and differential equations are more likely to top the list, at least judging by the number of pages these disciplines occupy in books on mathematical modelling in biology and health sciences.

Linear algebra is in many ways like the air we breathe. We might not appreciate it, but we panic once we realize it's not there any more. Without linear algebra, we would not know how to solve basic equations, how to work with vectors or matrices, or how to think about the geometry of the space that surrounds us. Linear algebra is about vectors, equations, lines, planes, matrices—it is everywhere:

* Tangent lines and tangent planes are linear objects; parametric equations of a line and a plane are important examples of parametric curves and surfaces.

* In some situations, solving for the critical points of a function of several variables involves solving a system of linear equations.

* Matrix multiplication is at the core of the *Markov chain,* an important probabilistic model that is often used in population dynamics (as well as in chemistry, physics, economics, game theory, and so on).

* The derivative (key concept!) of a function of several variables is a linear object, represented by a matrix.

* To find the *line of regression* (the line that best fits a given data set), we need to solve a system of linear equations.

* In order to solve a system of first-order linear differential equations, we need to find eigenvalues and eigenvectors, which are some of the essential tools in linear algebra.

* In many cases, when a math problem seems to be complicated, or difficult to handle, we *linearize* it to make it simpler, while still retaining important features.

Not to mention the thousands of applications...

In this module, we cover the basics of vectors, systems of linear equations, and matrices and linear transformations. We start by identifying the coordinate systems that we work with, describing different ways in which we can define the location of an object in a plane or in space. In Sections 2 and Section 3, we define vectors and study elementary vector operations.

We study various ways of describing lines and planes in Section 4. The geometry that we learn will help us understand the features of solutions of systems of linear equations, which we will start analyzing in Section 5. In Section 6, we learn about Gaussian elimination, an efficient technique for solving systems of equations. The remaining sections are devoted to matrices. We study basic properties and operations with matrices in Section 8, and in Section 9 we use inverse matrices to solve linear systems. We define linear transformations in Section 10, and in Section 11 we study eigenvectors and eigenvalues.

Two sections are entirely devoted to applications. In Section 7 we explore the way linear algebra is used to compute the image obtained from a CT scan, and in Section 12 we study the dynamics of population change based on a model that takes into account the dependence of reproduction on age.

The approach used in writing this module — clear explanations and easy-to-understand narratives; numerous graphs, simulations, pictures, and diagrams; a large number of fully solved examples and end-of-section exercises; and a wide spectrum of life sciences applications — makes the material suitable for students whose interests lie in life sciences and who are willing to deepen their understanding of life sciences phenomena.

I thank you for choosing this module, and I hope that you will like reading it and that you will learn some good and useful math.

Miroslav Lovrić
McMaster University, 2014

[Solutions to odd-numbered exercises from this module are posted (free download) on the Web page www.nelson.com/site/calculusforlifesciences.]

Outline

1 Identifying Location in a Plane and in Space

We construct **coordinate systems** that allow us to describe the location of a point in a plane or in space. Near the end of the section, we comment on spaces with many dimensions and on the meaning of coordinates in those spaces.

Cartesian Coordinate System in a Plane

A common way of identifying a location in a plane consists of using the (two-dimensional) *Cartesian coordinate system* (also known as the *xy-coordinate system,* or the *xy-plane*). The Cartesian coordinate system consists of two perpendicular number lines that intersect at the point that represents the number zero for both lines. This point is called the *origin* and is usually denoted by 0. The horizontal number line is called the *x-axis,* and the vertical line is the *y-axis;* the two coordinate axes divide the plane into four *quadrants,* as shown in Figure 1.1a.

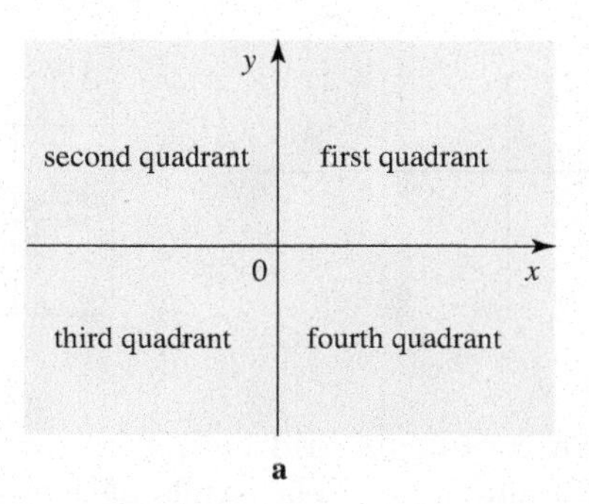

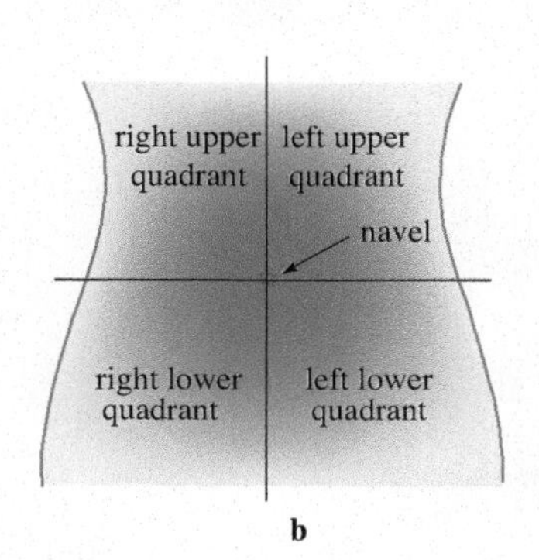

FIGURE 1.1
Cartesian coordinate system and application

The human abdominal area contains a large number of organs. In order to identify these organs, or to describe the location of injury or the source of pain, health professionals use the four quadrants shown in Figure 1.1b. In this case, the origin of the coordinate system is located at the umbilicus (navel), and the coordinate axes correspond to the horizontal and the vertical cross-sections of the body.

The word *two-dimensional* refers to the fact that there are two independent directions that could be referred to as left–right and up–down. The actual physical location of the origin of the coordinate system (as we have already seen, and will see in a number of examples) depends on the nature of the application.

A point A in the Cartesian coordinate system is uniquely determined by the following two real numbers: the directed distance from A to the y-axis, called the x-coordinate and denoted by x, and the directed distance from A to the x-axis, called the y-coordinate and denoted by y; see Figure 1.2a.

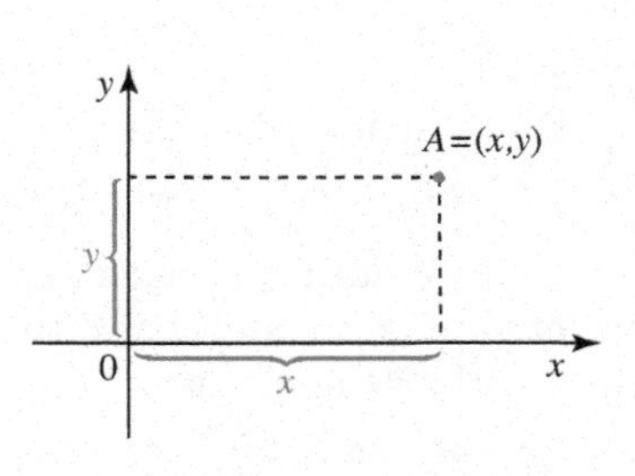

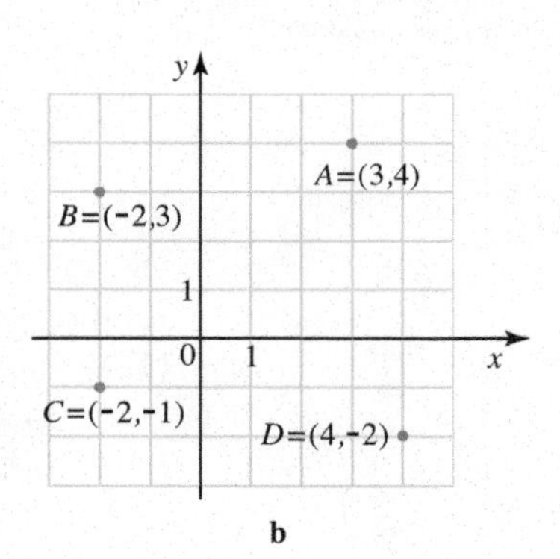

FIGURE 1.2
Cartesian coordinates

We use the notation $A = (x, y)$ to specify the coordinates of the point A. When we do not need to name the point, we just write (x, y). The word *directed* in *directed distance* means that the x-coordinates of points that lie to the left of the y-axis and the y-coordinates of points that lie below the x-axis are negative. Thus, both coordinates of the point A in Figure 1.2a are positive. The coordinates of the points in Figure 1.2b are $A = (3, 4)$, $B = (-2, 3)$, $C = (-2, -1)$, and $D = (4, -2)$.

A *coordinate grid* is obtained by drawing (usually equally spaced) lines parallel to the coordinate axes, such as in Figure 1.3a. The horizontal lines are given by $y = C$ and the vertical lines by $x = D$, where C and D are real numbers. Figure 1.3b shows a coordinate grid that was used in a study of navigation skills of desert ants. [The drawing is based on Merklea, T. & Wehnerc, R. (2010). Desert ants use foraging distance to adapt the nest search to the uncertainty of the path integrator. *Behavioral Ecology,* 21(2), 349–355.]

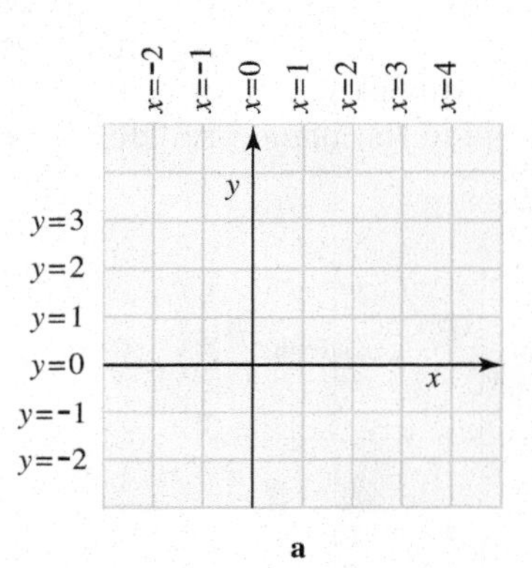

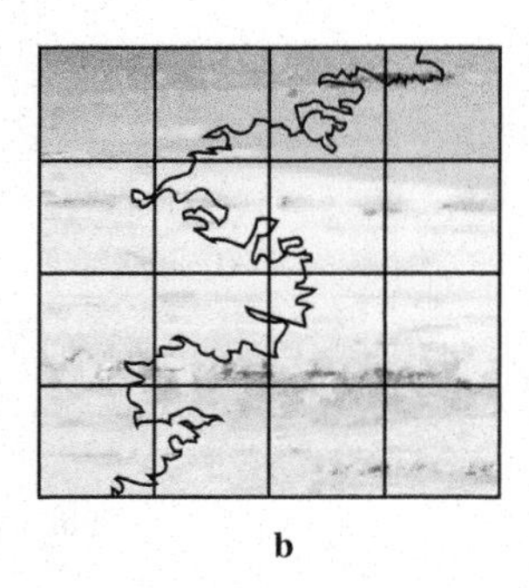

FIGURE 1.3

Coordinate grids

In Figure 1.4 a coordinate grid is used to compare the growth pattern in chimpanzee and human skulls. The skulls of newborn humans and chimpanzees are very similar. However, they develop in a different way, as indicated by the deformation of the coordinate grid. Note that the deformations did not preserve the distances (and so the grids representing adult skulls are no longer square or equally spaced).

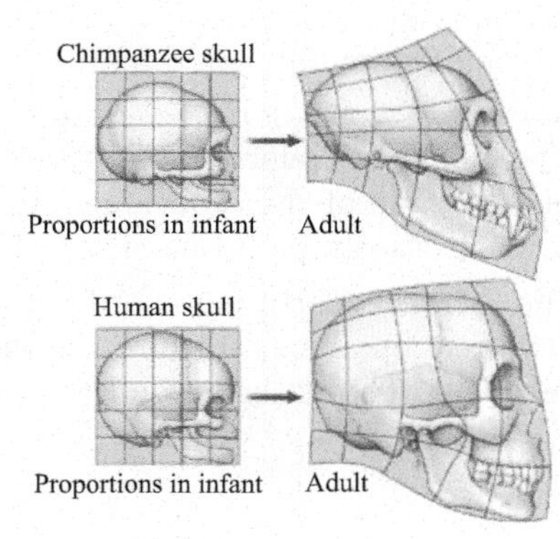

FIGURE 1.4

Growth pattern in human and chimpanzee skulls

(From RUSSELL/WOLFE/HERTZ/STARR. Biology, 1E.

We use the symbol $\mathbb{R}^2$ to denote the xy-plane:

$$\mathbb{R}^2 = \{(x, y) \mid x \in \mathbb{R} \text{ and } y \in \mathbb{R}\}$$

i.e., $\mathbb{R}^2$ is the set of all points (x, y) such that x and y are real numbers. The symbol (x, y) is an *ordered pair,* indicating that the order matters; for instance, $(1, 4)$ and $(4, 1)$ are not the same point.

A word of warning: the symbol $(1, 4)$ has multiple meanings: it could represent the coordinates of a point, or an interval of real numbers (as used in calculus), or the coordinates of a vector. We will make sure that the context is always clear, so that there will be no confusion about this.

Recall that the distance between two points $A = a$ and $B = b$ on a number line is given by $|b - a|$. Based on this, we obtain the formula for the distance $d(A, B)$ between points $A = (a_1, a_2)$ and $B = (b_1, b_2)$ in the xy-plane:

$$d(A, B) = \sqrt{(b_1 - a_1)^2 + (b_2 - a_2)^2}$$

How do we derive this formula? See Exercise 40.

Polar Coordinates in a Plane

One of the most fascinating examples of animal communication is the dance language of honey bees. Once she finds food, a forager bee returns to her colony with the samples of nectar or pollen. To tell other bees where the food is, she performs a dance on a vertical honeycomb (called the *tail-wagging dance,* or *waggle dance*) in order to communicate two pieces of information: the direction in which they should fly, given as an angle with respect to the sun, as well as the distance. (This discovery earned K. von Frisch the Nobel Prize in Physiology or Medicine in 1973.) Using this information, as well as odour cues (from the samples that the forager bee brought back), bees are able to locate the source of food.

FIGURE 1.5

Waggle dance

(From RUSSELL/WOLFE/HERTZ/STARR. Biology, 1E. © 2010 Nelson Education Ltd. Reproduced by permission. www.cengage.com/permissions)

Clearly, bees do not use the Cartesian coordinate system; instead, they use the *polar coordinate system,* which we now construct.

Pick a point in a plane (call it a *pole,* and denote it by 0) and a positive half-line (i.e., a number line containing 0 and all positive numbers) starting at 0; call that half-line the *polar axis;* see Figure 1.6a.

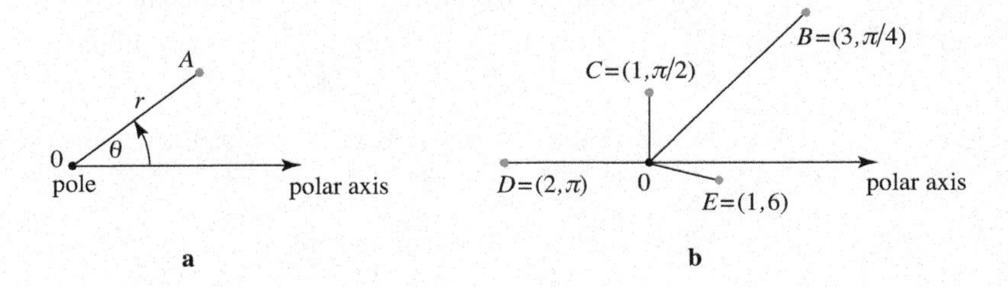

FIGURE 1.6

Polar coordinates in a plane

The location of a point A is uniquely determined by two numbers: the distance r ($r \geq 0$) from 0 to A, and the angle θ ($0 \leq \theta < 2\pi$) between the polar axis and the line segment $\overline{0A}$, measured counterclockwise from the polar axis; see Figure 1.6a.

To say that the polar coordinates of A are r and θ, we write $A = (r, \theta)$. All points in the plane have unique polar coordinates, except for the pole. The pole can be described as $0 = (0, \theta)$ for any angle $\theta, 0 \leq \theta < 2\pi$. The polar coordinates of the points in Figure 1.6b are $B = (3, \pi/4)$, $C = (1, \pi/2)$, $D = (2, \pi)$, and $E = (1, 6)$. (The coordinate $\theta = 6$ of the point E is 6 radians, i.e., $6(180/\pi) \approx 343.8^o$.)

The angle in polar coordinates does not have to be expressed in terms of π (look at the point E in Figure 1.6b). As well, if the coordinates of a point involve π, such as $(2, \pi)$, it does not necessarily mean that these are the polar coordinates of the point. To avoid confusion, we adopt the convention that whenever we use polar coordinates, we will explicitly say so; for instance, "the polar coordinates of A are $A = (3, 1)$." If the type of coordinates is not mentioned explicitly, then we assume that Cartesian coordinates (x, y) are being used.

To compare Cartesian and polar coordinates, we place the pole at the origin of the Cartesian system and the polar axis over the positive x-axis, as shown in Figure 1.7a.

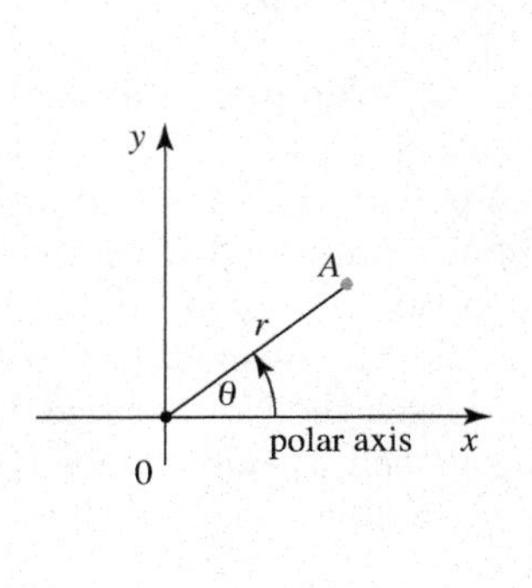

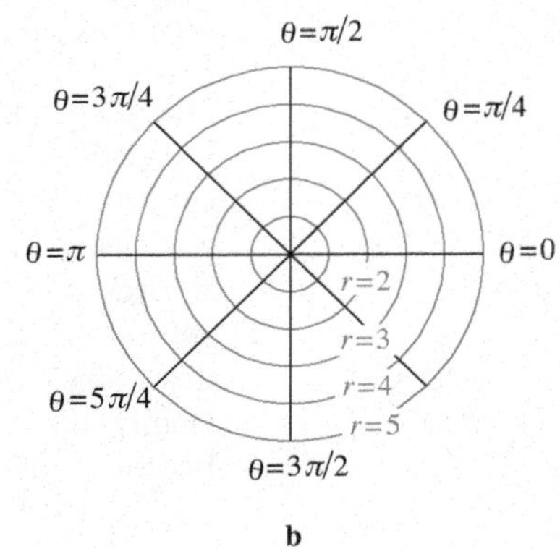

FIGURE 1.7

Polar coordinates within the Cartesian coordinate system and the polar coordinate grid

Thus, for all points on the positive x-axis, $\theta = 0$. The polar coordinates of points on the y-axis are $(r, \pi/2)$ if they lie above the x-axis and $(r, 3\pi/2)$ if they lie below it (keep in mind that $r \geq 0$). For points on the negative x-axis, $\theta = \pi$.

The equation $\theta = C$, where C is a fixed real number $(0 \leq C < 2\pi)$, describes all points on a half-line starting at the pole that make an angle of C radians (measured counterclockwise) with respect to the polar axis (i.e., the x-axis). The equation $r = D$ (where $D > 0$) describes all points with polar coordinates (D, θ), where D is a fixed number and $0 \leq \theta < 2\pi$. In other words, $r = D$ is the set of points whose distance from the pole is equal to D; it is a circle of radius D centred at the pole.

The coordinate grid in polar coordinates consists of the half-lines emanating from the pole and concentric circles centred at the pole; see Figure 1.7b.

We use Figure 1.8 to derive the conversion formulas between the Cartesian and the polar coordinate systems. From $\cos\theta = x/r$ and $\sin\theta = y/r$ we obtain

$$\begin{aligned} x &= r\cos\theta \\ y &= r\sin\theta \end{aligned} \tag{1.1}$$

Given the Cartesian coordinates (x, y), we use the Pythagorean theorem to obtain

$$r = \sqrt{x^2 + y^2} \tag{1.2}$$

The angle θ is determined from

$$\tan\theta = \frac{y}{x} \tag{1.3}$$

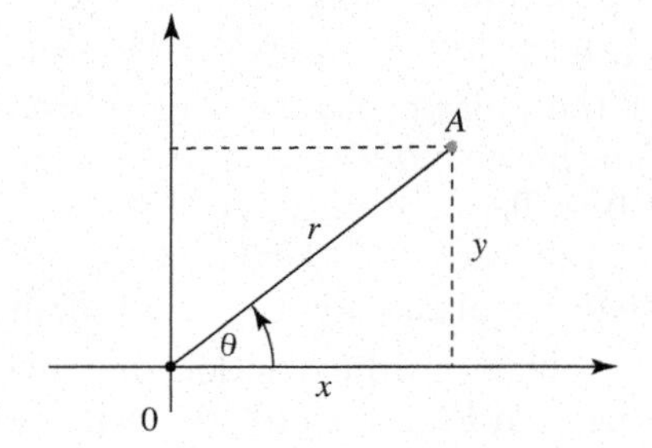

FIGURE 1.8

Cartesian and polar coordinates

Example 1.1 Conversion between Polar and Cartesian Coordinates

If $A = (3, \pi/6)$ are the polar coordinates of A, then its Cartesian coordinates are

$$A = \big(3\cos(\pi/6), 3\sin(\pi/6)\big) = \left(\frac{3\sqrt{3}}{2}, \frac{3}{2}\right)$$

If, in polar coordinates, $B = (2, 3\pi/2)$, then

$$B = \big(2\cos(3\pi/2), 2\sin(3\pi/2)\big) = (0, -2)$$

in Cartesian coordinates. If $C = (4, 3)$ in polar coordinates, then the Cartesian coordinates of C are

$$B = \big(4\cos 3, 4\sin 3\big) \approx (-3.96, 0.56)$$

Assume that $D = (1, 1)$ in Cartesian coordinates. Then

$$r = \sqrt{1^2 + 1^2} = \sqrt{2}$$

and from

$$\tan\theta = \frac{1}{1} = 1$$

we conclude that $\theta = \pi/4$. Thus, $D = (\sqrt{2}, \pi/4)$ in polar coordinates. If $E = (-3, 3)$ in Cartesian coordinates, then

$$r = \sqrt{(-3)^2 + 3^2} = \sqrt{18} = 3\sqrt{2}$$

and

$$\tan\theta = \frac{3}{-3} = -1$$

We conclude that $\theta = -\pi/4 + k\pi$, where k is an integer. Because the point $(-3, 3)$ is in the second quadrant, we must pick $\theta = -\pi/4 + 2\pi = 3\pi/4$. Thus, the polar coordinates of E are $E = (3\sqrt{2}, 3\pi/4)$. ▲

Note that in (1.3) we said $\tan\theta = y/x$, and not $\theta = \arctan(y/x)$. Why? Because the ranges of θ do not agree: if θ is a polar coordinate, then $0 \leq \theta < 2\pi$, whereas if $\theta = \arctan(y/x)$, then $-\pi/2 < \theta < \pi/2$. That is why we needed to adjust the angle in calculating the coordinates of the point E in the previous example. To gain more experience with this, look at Exercises 26 to 37.

Three-Dimensional Cartesian Coordinate System

The *three-dimensional Cartesian coordinate system* consists of three mutually perpendicular number lines that intersect at the point 0 (called the *origin*) that represents the number 0 for all of them.

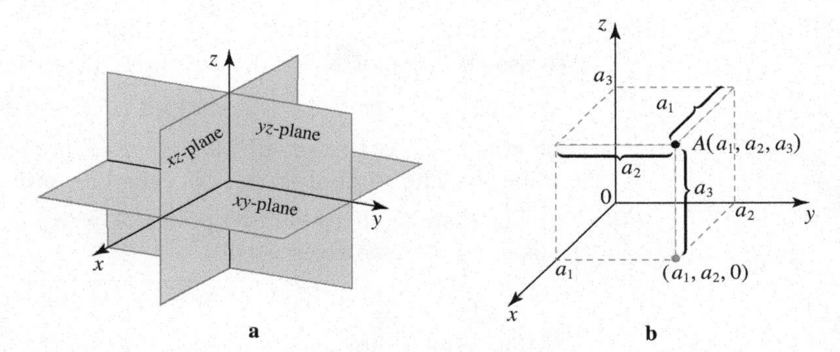

FIGURE 1.9

The three coordinate planes and the Cartesian coordinates of a point

The three number lines, called the *coordinate axes,* are identified as the x-axis, the y-axis, and the z-axis. The pairs of coordinate lines form the *coordinate planes:* the xy-plane, the yz-plane, and the xz-plane; see Figure 1.9a.

The location of a point A in space can be uniquely identified using three numbers: the directed distance a_1 from A to the yz-plane, called the x-coordinate of A; the directed distance a_2 from A to the xz-plane, called the y-coordinate of A; and the directed distance a_3 from A to the xy-plane, called the z-coordinate of A (see Figure 1.9b). "Directed distance" means the usual distance, to which we assign the sign $+$ or $-$ depending on the location of A, according to Table 1.1.

Table 1.1

The location of a point	The sign of its coordinate
above the xy-plane	z-coordinate is positive
below the xy-plane	z-coordinate is negative
right of the xz-plane	y-coordinate is positive
left of the xz-plane	y-coordinate is negative
in front of the yz-plane	x-coordinate is positive
behind the yz-plane	x-coordinate is negative

We write $A = (a_1, a_2, a_3)$ and say that the point A has coordinates (a_1, a_2, a_3). The origin has coordinates $(0, 0, 0)$.

In studying human anatomy, we refer to the three coordinate planes as *sagittal* (median vertical plane, which separates the left half from the right half), *transverse* (horizontal plane, which separates the top half from the bottom half) and *coronal* (frontal vertical plane, which separates the front half from the back half).

Sometimes, the coordinate planes are referred to by the three independent directions shown in Figure 1.10.

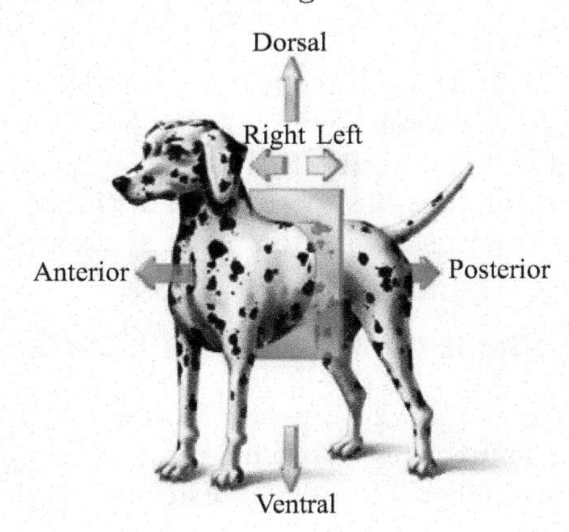

FIGURE 1.10

Coordinate directions as used in anatomy

(From RUSSELL/WOLFE/HERTZ/STARR. Biology, 1E.

We use the symbol $\mathbb{R}^3$ to denote three-dimensional space:

$$\mathbb{R}^3 = \{(x, y, z) \mid x \in \mathbb{R},\ y \in \mathbb{R}, \text{ and } z \in \mathbb{R}\}$$

The space $\mathbb{R}^3$ is the set of all points (x, y, z) all of whose coordinates are real numbers. (The symbol (x, y, z) is called an *ordered triple.*)

The distance between two points $A = (a_1, a_2, a_3)$ and $B = (b_1, b_2, b_3)$ in space $\mathbb{R}^3$ is given by (see Exercise 40)

$$d(A, B) = \sqrt{(b_1 - a_1)^2 + (b_2 - a_2)^2 + (b_3 - a_3)^2}$$

The coordinate axes divide the plane $\mathbb{R}^2$ into four quadrants. Moving one dimension higher, we say that the three coordinate planes divide the space $\mathbb{R}^3$ into eight octants, defined (as with the quadrants in $\mathbb{R}^2$) by the signs of the coordinates, as shown in Table 1.2.

Table 1.2

x	y	z	Location	Octant
+	+	+	top-front-right	1st
−	+	+	top-back-right	2nd
−	−	+	top-back-left	3rd
+	−	+	top-front-left	4th
+	+	−	bottom-front-right	5th
−	+	−	bottom-back-right	6th
−	−	−	bottom-back-left	7th
+	−	−	bottom-front-left	8th

Higher Dimensions and the Meaning of Coordinates

Generalizing the definitions of the plane $\mathbb{R}^2$ and the space $\mathbb{R}^3$, we obtain the *n-dimensional space*

$$\mathbb{R}^n = \{(x_1, x_2, \ldots, x_n) \mid x_1 \in \mathbb{R},\ x_2 \in \mathbb{R},\ \ldots,\ x_n \in \mathbb{R}\}$$

A point in $\mathbb{R}^n$ is uniquely represented by an *ordered n-tuple* $(x_1, x_2, \ldots, x_n)$.

Why do we need more than three dimensions? Our discussion, so far, has implied that all coordinates are given the same meaning (distance) and use the same units. However, in numerous situations this will not be so. For instance, when we study interactions between two species (say, rabbits and foxes), we use a coordinate plane with axes representing rabbits and foxes. The ordered pair $(107, 12)$, plotted as the point with coordinates 107 and 12, represents the moment when there are 107 rabbits and 12 foxes. (Note that, in this context, the distance between two points — although we can compute it — might not be useful; see Exercise 43.)

Studying age-related population changes in some country (or in some region), researchers might be interested in tracking the changes within, say, six age groups: newborn to 9, 10–19, 20–29, 30–39, 40–49, and 50–59 years old. In this case, the ordered 6-tuple $(0.2, 0.25, 0.3, 0.15, 0.08, 0.02)$, i.e., an element of a six-dimensional space, is used to record the fact that 20% of the population is in the newborn to 9 years age group, 25% of the population is in the 10–19 age group, 30% of the population falls in the 20–29 age group, and so on.

Quite often, the coordinates of a point are related to the variables in a system of linear equations. A solution of a system of ten equations with ten variables can be represented as a point in ten-dimensional space. Describing the mathematics of computed tomography (CT) scanning, we will see how one can easily obtain points that belong to spaces of more than ten thousand dimensions.

In certain situations (calculus and vector calculus and their applications) the coordinates of a point are functions rather than real numbers.

A colour (for instance for printing purposes) is computer-coded as an ordered quadruple (c, m, y, k); the coordinates represent the amounts of cyan, magenta, yellow, and black that are mixed to produce a desired colour. For instance, the darker blue colour in Table 1.2 is defined by the coordinates $(100, 44, 0, 0)$, and the coordinates of the lighter shade of blue are $(7, 3, 0, 0)$.

Remark Linear algebra in two or three dimensions does not differ much from linear algebra in many-dimensional spaces. For that reason, we focus on two and three dimensions, and then generalize if needed.

Summary In this section we constructed the **plane** $\mathbb{R}^2$ and the **three-dimensional space** $\mathbb{R}^3$, and mentioned a general **n-dimensional space** $\mathbb{R}^n$. The location of a point in a plane can be described using **Cartesian coordinates** or **polar coordinates.** In general, the coordinates of a point can represent a wide variety of objects, such as the solutions of a system of equations, the age distribution within a population, or the number of members of interacting species in an ecosystem.

1 Exercises

1. Which of the three points $A = (1, 3)$, $B = (0, 4)$, and $C = (2, 2)$ is closest to the origin?
2. Which of the three points $A = (-4, 2)$, $B = (0, 1)$, and $C = (2, 0)$ is closest to the point $(5, 0)$?
3. Find a so that the distance between the points $(3, -1)$ and $(a, 1)$ is 10.
4. Find the value of a so that the distance between the points $(1, 1)$ and $(2, a)$ is 12.
5. Find the distance between the points $(1, 2, 3)$ and $(4, 5, 6)$ in $\mathbb{R}^3$.
6. Which of the two points $(2, -1, -3)$ and $(4, 0, 1)$ is closer to the origin?
7. Which of the two points $(3, -1, 2)$ and $(0, 0, 6)$ is closer to the point $(9, 0, -4)$?
8. Plot the points whose polar coordinates are $(2, \pi/4)$, $(3, 3\pi/4)$, and $(4, 5\pi/4)$.
9. Plot the points whose polar coordinates are $(1, \pi/4)$, $(1, \pi/2)$, and $(1, 1)$.

10–15 ▪ Convert the polar coordinates of each point to the corresponding Cartesian coordinates.

10. $(2, \pi/4)$
11. $(10, \pi)$
12. $(4, 5\pi/4)$
13. $(10, 3\pi/2)$
14. $(12, 3)$
15. $(4, 4)$

16. Find the distance between the points whose polar coordinates are $(2, \pi/6)$ and $(2, \pi/3)$.
17. Find the distance between the points whose polar coordinates are $(4, \pi/4)$ and $(1, 3\pi/4)$.

18–23 ▪ Convert the Cartesian coordinates of each point to the corresponding polar coordinates.

18. $(2, 2)$
19. $(-5, 5)$
20. $(\sqrt{3}, 1)$
21. $(1, \sqrt{3})$
22. $(7, 1)$
23. $(-2, 5)$

24. What curve do the points with polar coordinates $(3, \pi/4)$, $(3, 7\pi/6)$, and $(3, 5)$ belong to?
25. What curve do the points with polar coordinates $(1, 2)$, $(2, 2)$, and $(3, 2)$ belong to?

26–31 ▪ In each case find the angle θ (with $0 \le \theta < 2\pi$).

26. $\tan\theta = \sqrt{3}$; θ is in the third quadrant
27. $\tan\theta = -\sqrt{3}$; θ is in the second quadrant
28. $\tan\theta = -\sqrt{3}$; θ is in the fourth quadrant
29. $\tan\theta = -1$; θ is in the fourth quadrant
30. $\tan\theta = 1$; θ is in the third quadrant
31. $\tan\theta = -1$; θ is in the second quadrant

32–37 ▪ Convert the Cartesian coordinates of each point to its polar coordinates. Make sure that you identify the correct angle θ.

32. $(-1,-1)$

33. $(-1,1)$

34. $(1,-1)$

35. $(-1,-\sqrt{3})$

36. $(1,-\sqrt{3})$

37. $(-\sqrt{3},-1)$

38. Show that the point $(4,2,-1)$ is located halfway between the points $(3,0,7)$ and $(5,4,-9)$.

39. Find a so that the point $(a,4)$ is halfway between the points $(-2,3)$ and $(3,5)$.

40. Using the given diagram and the Pythagorean Theorem, derive the formula

$$d(A,B)=\sqrt{(b_1-a_1)^2+(b_2-a_2)^2}$$

for the distance between the points $A=(a_1,a_2)$ and $B=(b_1,b_2)$. Using the Pythagorean Theorem again (draw a diagram!), derive the formula for the distance

$$d(A,B)=\sqrt{(b_1-a_1)^2+(b_2-a_2)^2+(b_3-a_3)^2}$$

between the points $A=(a_1,a_2,a_3)$ and $B=(b_1,b_2,b_3)$ in three-dimensional space.

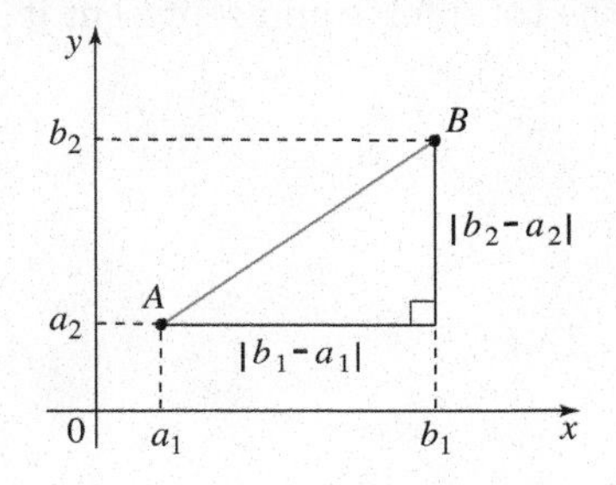

41. Prove that the point with coordinates $((a_1+b_1)/2,(a_2+b_2)/2)$ is halfway between the points (a_1,a_2) and (b_1,b_2).

42. Show that the distance between two points in a plane whose polar coordinates are $A=(r_1,\theta_1)$ and $B=(r_2,\theta_2)$ is given by

$$d(A,B)=\sqrt{r_1^2+r_2^2-2r_1r_2\cos(\theta_1-\theta_2)}$$

43. Last year's population of 107 rabbits and 12 foxes is represented by the point $(107,12)$ in the rabbits – foxes coordinate system. What can you conclude about the change in the number of rabbits and foxes if the current year's population is represented by a point whose distance from $(107,12)$ is less than 5? Less than 50? More than 100?

2 Vectors

In this section, we introduce **vectors** in a plane and in space, study their properties (such as **length**), and define elementary operations (**addition of vectors** and **multiplication of a vector by a scalar**). As an important object involving these operations, we define the **linear combination** of vectors.

Vectors in a Plane

Figure 2.1 shows a vertical cross-section of the tibia (shin bone) and the forces that act on or near its top during robust motion (stop-jump) of the knee. [The diagram is adapted from Chappell, J.D., Garrett, W.E., & Yu, B. (2006). Authors' Response: Effect of fatigue on knee kinetics and kinematics in stop-jump tasks. *The American Journal of Sports Medicine,* 34(2), 313–315.]

The forces are shown as vectors anchored at different locations on the tibia. For various reasons, it is important to know what the total force exerted on the top of the tibia is. How can we calculate the total force and represent it geometrically?

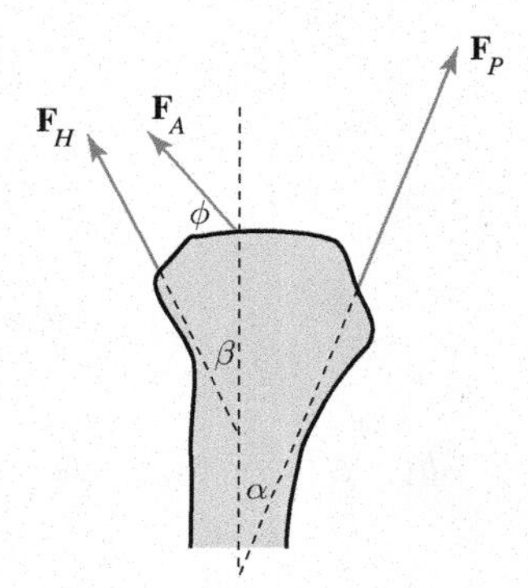

FIGURE 2.1
Forces acting on the tibia

Later in this section, when we are ready to answer this question, we will name the forces involved and give the data that we need. First, we define what a vector is and introduce elementary operations with vectors.

Definition 1 **Vector in a Plane**

A *two-dimensional vector* (or a *vector* in $\mathbb{R}^2$) is an ordered pair $\mathbf{v} = [v_1, v_2]$ of real numbers.

To distinguish between a point and a vector, we use square brackets for vectors. Written in the form $\mathbf{v} = [\,v_1 \quad v_2\,]$, the vector $\mathbf{v}$ is also called a *row vector.* Written as $\mathbf{v} = \begin{bmatrix} v_1 \\ v_2 \end{bmatrix}$, it is a *column vector.* (Which form of vector (ordered pair, row, or column) we use depends on the context.) The real numbers v_1 and v_2 are the *components* of $\mathbf{v}$. Instead of using boldface symbols to denote vectors, sometimes we place an arrow over the symbol for a vector and write $\overrightarrow{v} = [v_1, v_2]$.

Next, we introduce a technical object that will help us understand vectors a bit better, as well as facilitate working with them.

Definition 2 **Directed Line Segment**

A *directed line segment* $\overrightarrow{AB}$ is a line segment $\overline{AB}$ equipped with a direction ("from A to B": the point A is its *initial point* (or *tail,* also called a *footpoint*), and B is its *terminal point* (or *head,* also called a *tip*).

We visualize a vector $\mathbf{v} = [v_1, v_2]$ as the directed line segment from the origin 0 to the point $V = (v_1, v_2)$. Sometimes we will refer to $\mathbf{v}$ as the *vector from* 0 *to* V or the *displacement from* 0 *to* V; see Figure 2.2.

As we will soon see, there are many directed line segments corresponding to the same vector.

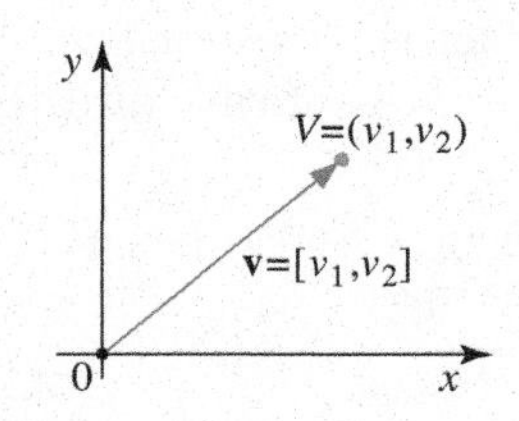

FIGURE 2.2

A vector represented by a directed line segment

Points are ordered pairs, and so are vectors. So what is $\mathbb{R}^2$? From what we said, it follows that we can view $\mathbb{R}^2$ *either as a set of points or as a set of vectors.* The choice depends on the context within which we work. It is easy to switch between the two: a point $V = (v_1, v_2)$ in $\mathbb{R}^2$ is viewed as the vector $[v_1, v_2]$ whose tail is at the origin and whose head is the point (v_1, v_2). Likewise, to each vector $\mathbf{v} = [v_1, v_2]$ we associate the point (v_1, v_2), that is its head; see Figure 2.2. The vector $[v_1, v_2]$ is called the *position vector* of the point (v_1, v_2).

We have already mentioned that there is a difference in notation: for points we use round brackets; whereas square brackets are used to denote vectors.

The vector as we defined it (i.e., with its tail at the origin) is called a position vector or a *vector in standard position.* For many reasons, we would like to have vectors that can be *moved around,* i.e., whose tails can be located anywhere, not necessarily at the origin. This is why we need directed line segments.

Let $\mathbf{v} = [v_1, v_2]$ be a vector (with tail at the origin) and pick a point $A = (a_1, a_2)$. The *representative of the vector* $\mathbf{v}$ *with initial point* A is the directed line segment $\overrightarrow{AB}$, where

$$B = (a_1 + v_1, a_2 + v_2) \tag{2.1}$$

See Figure 2.3.

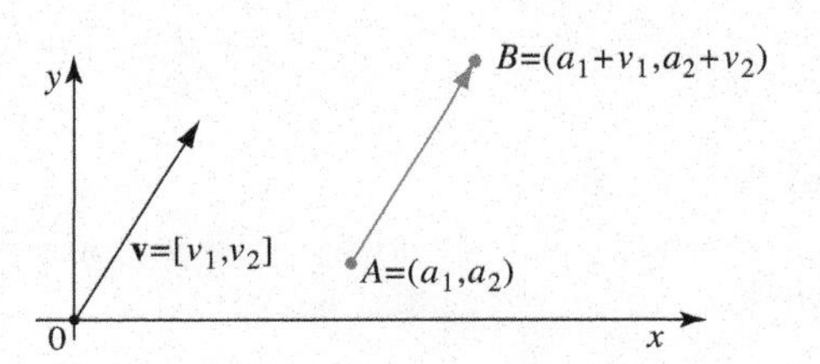

FIGURE 2.3

Directed line segment representing a vector

For example, let $\mathbf{v} = [2, 3]$. The representative of $\mathbf{v}$ that starts at the point $(4, 0)$ is the directed line segment from $(4, 0)$ to $(4+2, 0+3) = (6, 3)$; see Figure 2.4a. The representative of the vector $\mathbf{v}$ that starts at $(-3, -2)$ is the directed line segment from $(-3, -2)$ to $(-1, 1)$. All directed line segments in Figure 2.4b represent the vector $\mathbf{v} = [2, 3]$.

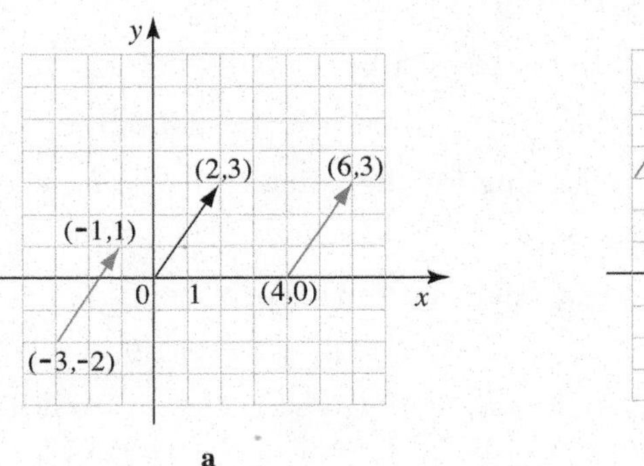

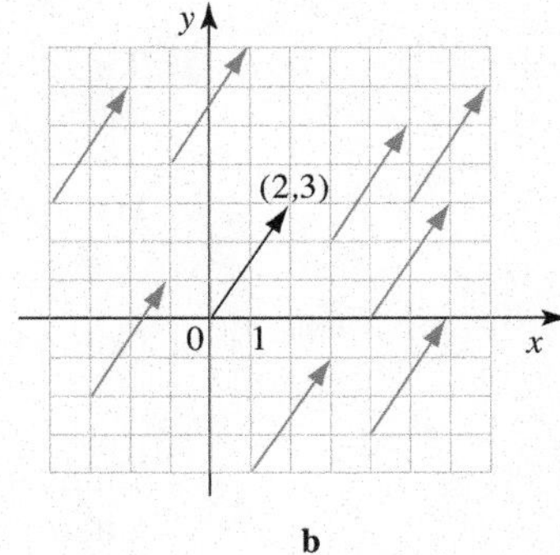

FIGURE 2.4

Directed line segments representing the same vector

Thus, we think of a vector as an infinite collection of directed line segments (one starting at each point in $\mathbb{R}^2$) that are parallel to each other, have the same length, and point in the same direction. In many situations (especially in applications) we blur the distinction between vectors and directed line segments and refer to all of them as vectors.

The vector $\mathbf{0} = [0,0]$, both of whose components are 0, is called the *zero vector.* We visualize the zero vector as a directed line segment $\overrightarrow{AA}$ that starts and ends at the same point; thus, all representatives of the zero vector are points.

Take two points, $A = (a_1, a_2)$ and $B = (b_1, b_2)$, and construct the "vector from A to B"; to be precise, we would like to find the vector whose representative directed line segment is $\overrightarrow{AB}$.

Let $\mathbf{v} = [v_1, v_2]$. Combining $B = (b_1, b_2)$ and $B = (a_1 + v_1, a_2 + v_2)$ (see (2.1)) we get $b_1 = a_1 + v_1$ and $b_2 = a_2 + v_2$, and thus

$$\mathbf{v} = [v_1, v_2] = [b_1 - a_1, b_2 - a_2] \tag{2.2}$$

So the coordinates of $\mathbf{v}$ are the differences ("head minus tail") of the respective coordinates of A and B. We also say that the displacement $\mathbf{v}$ is "the terminal point minus the initial point."

Example 2.1 A Vector and Its Representative Directed Line Segments

Show that the directed line segments $\overrightarrow{AB}$ and $\overrightarrow{CD}$, where $A = (-2, 4)$, $B = (4, 0)$, $C = (3, 2)$, and $D = (9, -2)$, represent the same vector.

▶ The directed line segment $\overrightarrow{AB}$ represents the vector ("terminal point minus initial point")

$$[4 - (-2), 0 - 4] = [6, -4]$$

The directed line segment $\overrightarrow{CD}$ represents the same vector, since

$$[9 - 3, -2 - 2] = [6, -4]$$

▲

Definition 3 Equal Vectors

Two vectors $\mathbf{v} = [v_1, v_2]$ and $\mathbf{w} = [w_1, w_2]$ are said to be *equal* if their components are equal, i.e., if $v_1 = w_1$ and $v_2 = w_2$. ▲

For instance, from $[v_1, v_2] = [3, -5]$ we conclude that $v_1 = 3$ and $v_2 = -5$.

Example 2.2 Another Vector and Its Representative Directed Line Segments

A vector $\mathbf{v}$ is represented by the directed line segment $\overrightarrow{AB}$, where $A = (2.3, -1.1)$ and $B = (0.4, 3.9)$. Find the representative directed line segment of $\mathbf{v}$ that starts at the point $C = (-3, -0.7)$.

▶ The coordinates of $\mathbf{v}$ are

$$\mathbf{v} = [0.4 - 2.3, 3.9 - (-1.1)] = [-1.9, 5]$$

We are looking for the point $D = (d_1, d_2)$ such that $\mathbf{v}$ is represented by $\overrightarrow{CD}$, i.e.,

$$\mathbf{v} = [d_1 - (-3), d_2 - (-0.7)] = [d_1 + 3, d_2 + 0.7] = [-1.9, 5]$$

Comparing the coordinates of $\mathbf{v}$ yields

$$d_1 + 3 = -1.9 \quad \text{and} \quad d_2 + 0.7 = 5$$

Thus, $d_1 = -4.9$ and $d_2 = 4.3$; the directed line segment we are looking for is $\overrightarrow{CD}$, where $C = (-3, -0.7)$ and $D = (-4.9, 4.3)$. ▲

Example 2.3 Vector as Displacement

If $A = (-3, 2)$ and $B = (1, -4)$, then

$$\mathbf{v} = [1 - (-3), -4 - 2] = [4, -6]$$

is the displacement vector from A to B. The displacement vector from B to A is

$$\mathbf{w} = [-3 - 1, 2 - (-4)] = [-4, 6]$$

Note that the respective components of $\mathbf{v}$ and $\mathbf{w}$ are of opposite signs.

Definition 4 Length of a Vector

The *length* (also called *magnitude,* or *norm*) of a vector $\mathbf{v} = [v_1, v_2]$ is given by

$$\|\mathbf{v}\| = \sqrt{v_1^2 + v_2^2}$$

For example, the length of the vector $\mathbf{v} = [3, -5]$ is

$$\|\mathbf{v}\| = \sqrt{(3)^2 + (-5)^2} = \sqrt{34}$$

Assume that $\mathbf{v}$ is represented by the directed line segment $\overrightarrow{AB}$, with $A = (a_1, a_2)$ and $B = (b_1, b_2)$. Then $\mathbf{v} = [b_1 - a_1, b_2 - a_2]$, and the length of $\mathbf{v}$

$$\|\mathbf{v}\| = \sqrt{(b_1 - a_1)^2 + (b_2 - a_2)^2}$$

is equal to the distance between A and B.

For example, if $\mathbf{v}$ is represented by $\overrightarrow{AB}$, with $A = (2, -7)$ and $B = (-3, 5)$, then $\mathbf{v} = [-5, 12]$ and

$$\|\mathbf{v}\| = \sqrt{(-5)^2 + (12)^2} = \sqrt{169} = 13$$

The distance between A and B is 13 (units).

Note that $\|\mathbf{v}\| \geq 0$ for all vectors $\mathbf{v}$.

The length of the zero vector is $\|\mathbf{0}\| = \sqrt{0^2 + 0^2} = 0$. If $\mathbf{v} = [v_1, v_2]$ is a vector such that $\|\mathbf{v}\| = 0$, then

$$\|\mathbf{v}\| = \sqrt{v_1^2 + v_2^2} = 0$$

implies that $v_1^2 + v_2^2 = 0$; i.e., $v_1 = 0$ and $v_2 = 0$, and so $\mathbf{v} = \mathbf{0}$. To summarize: the zero vector is of length zero and is the only vector with that property. Formally, we say that $\|\mathbf{v}\| = 0$ if and only if $\mathbf{v} = \mathbf{0}$.

Example 2.4 Direction and Magnitude of a Vector in Applications

Figure 2.5a shows the blood velocity profile in a cross-section of a blood vessel. We use vectors to indicate both the speed and the direction of the blood flowing through the vessel. Blood flows most quickly at the centre of the vessel. Near the wall of the vessel, its speed is small.

In Figure 2.5b we illustrate the process of diffusion out through a cell membrane. Vectors are used to indicate the direction of diffusion, as well as its intensity.

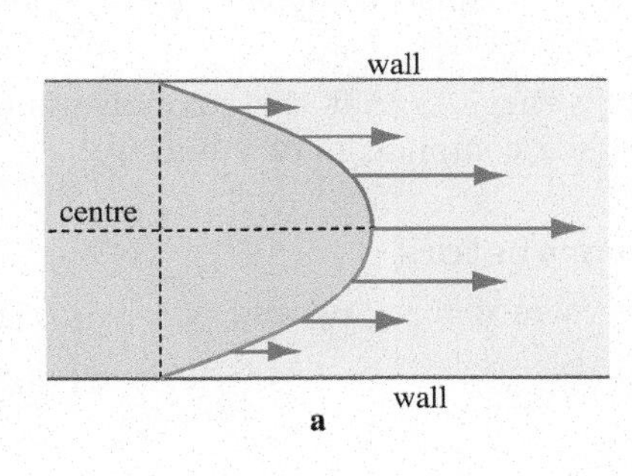

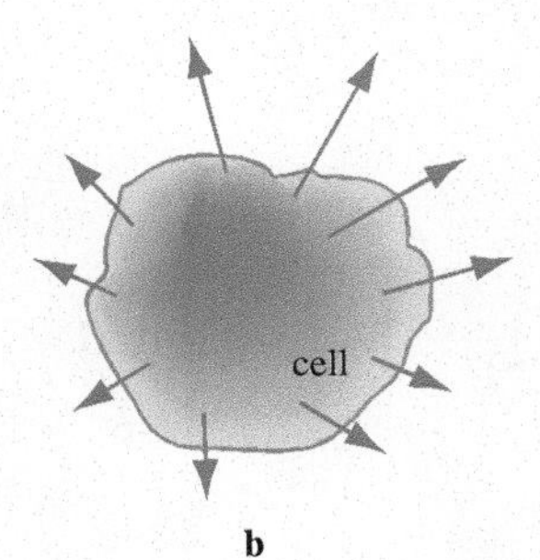

FIGURE 2.5

Using vectors to describe blood flow and diffusion through a membrane

Definition 5 Unit Vector

A vector whose length is equal to 1 is called a *unit vector.*

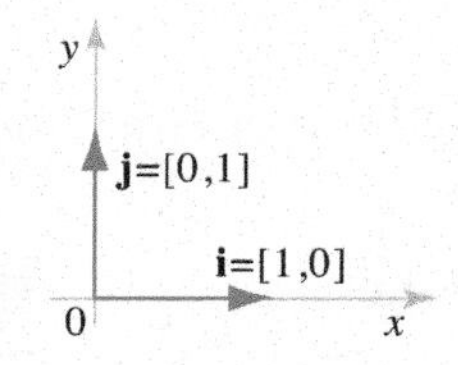

FIGURE 2.6
Standard unit vectors

For instance, the vector $\mathbf{v} = [4/5, 3/5]$ is a unit vector because

$$\|\mathbf{v}\| = \sqrt{\left(\frac{4}{5}\right)^2 + \left(\frac{3}{5}\right)^2} = \sqrt{1} = 1$$

In $\mathbb{R}^2$, there are two special unit vectors, called the *standard unit vectors* or *coordinate (unit) vectors,* given by $\mathbf{i} = [1, 0]$ and $\mathbf{j} = [0, 1]$; see Figure 2.6.

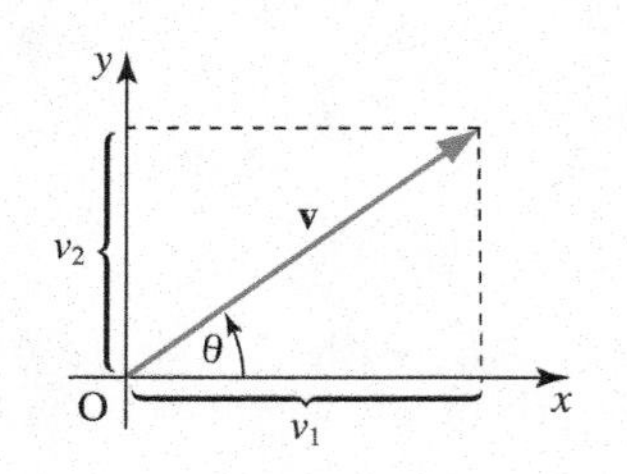

FIGURE 2.7
The polar form of a vector

Consider a non-zero vector $\mathbf{v} = [v_1, v_2]$ in standard position, i.e., represented by the directed line segment from the origin to the point (v_1, v_2). Denote by θ the angle (measured counterclockwise) between the positive x-axis and the vector $\mathbf{v}$. Then (see Figure 2.7) $\cos\theta = v_1/\|\mathbf{v}\|$ and $\sin\theta = v_2/\|\mathbf{v}\|$, and therefore

$$v_1 = \|\mathbf{v}\| \cos\theta$$
$$v_2 = \|\mathbf{v}\| \sin\theta$$

(We assumed that $\mathbf{v}$ is a non-zero vector, so that $\|\mathbf{v}\| \neq 0$; this fact allowed us to divide by $\|\mathbf{v}\|$.)

We have discovered the *polar form* of a vector:

$$\mathbf{v} = [\|\mathbf{v}\| \cos\theta, \|\mathbf{v}\| \sin\theta] \tag{2.3}$$

Note that (2.3) holds for $\mathbf{v} = \mathbf{0}$ as well.

Example 2.5 Polar Form of a Vector

Find a vector $\mathbf{v}$ in the plane whose length is 11 and whose direction makes an angle of $\pi/6$ radians (measured counterclockwise) with respect to the positive x-axis.

▶ We are given that $\|\mathbf{v}\| = 11$ and $\theta = \pi/6$. The polar form of $\mathbf{v}$ is

$$\begin{aligned}\mathbf{v} &= [\|\mathbf{v}\| \cos\theta, \|\mathbf{v}\| \sin\theta] \\ &= [11\cos(\pi/6), 11\sin(\pi/6)] = \left[11\sqrt{3}/2, 11/2\right]\end{aligned}$$

Vector Operations

First, we define how to add vectors $\mathbf{v} = [v_1, v_2]$ and $\mathbf{w} = [w_1, w_2]$.

Think of the two vectors as displacements: $\mathbf{v}$ displaces an object from the origin 0 to the point B, and $\mathbf{w}$ further displaces it from B to C; see Figure 2.8a. In the language of directed line segments, we pick a point (the origin, for convenience), represent $\mathbf{v}$ by the directed line segment $\overrightarrow{0B}$, and then represent $\mathbf{w}$ by the directed line segment $\overrightarrow{BC}$ whose initial point coincides with the terminal point of $\overrightarrow{0B}$.

Looking at Figure 2.8a we see that the net (total) displacement is given by the vector (directed line segment) whose tail is at 0 and whose head is at C. We say that the total displacement is given by the *sum* $\mathbf{v} + \mathbf{w}$ of the two vectors $\mathbf{v}$ and $\mathbf{w}$.

Figure 2.8b suggests that the components of $\mathbf{v} + \mathbf{w}$ are obtained by adding the corresponding components of $\mathbf{v}$ and $\mathbf{w}$.

Definition 6 Sum of Two Vectors

The sum of vectors $\mathbf{v} = [v_1, v_2]$ and $\mathbf{w} = [w_1, w_2]$ is the vector

$$\mathbf{v} + \mathbf{w} = [v_1 + w_1, v_2 + w_2]$$

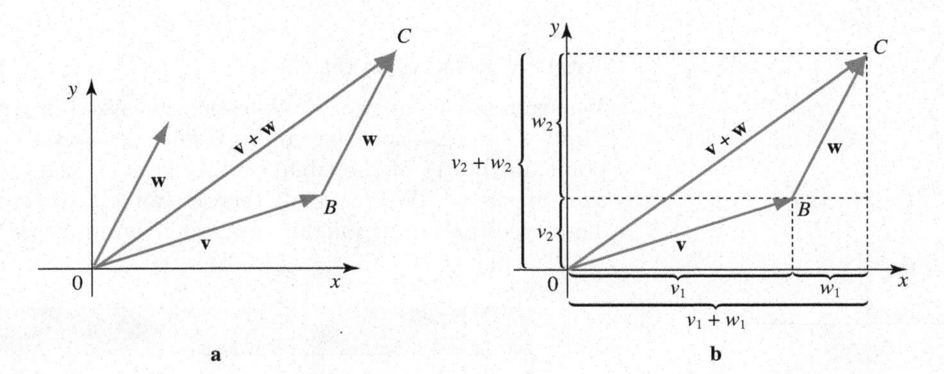

FIGURE 2.8

Adding two vectors

The formula in Definition 6 tells us how to calculate the sum of two vectors algebraically (i.e., when their components are known). Our geometric reasoning at the start of this subsection (Figure 2.8a) gives the *triangle law* for adding vectors.

Alternatively, we could use the *parallelogram law:* view **v** and **w** in standard position (with their tails at the origin) and form the parallelogram defined by **v** and **w**; see Figure 2.9. The diagonal of the parallelogram that starts at the origin and ends at the opposite vertex is the sum **v** + **w**.

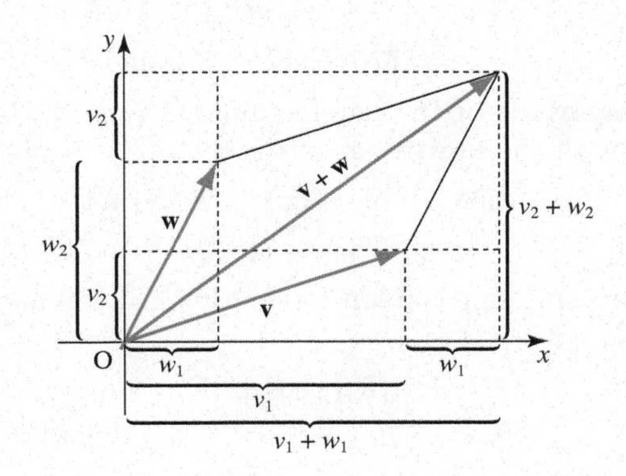

FIGURE 2.9

Adding vectors using the parallelogram law

Example 2.6 Adding Vectors

The sum of the vectors $\mathbf{v} = [-2, 2]$ and $\mathbf{w} = [4, 3]$ is the vector

$$\mathbf{v} + \mathbf{w} = [-2, 2] + [4, 3] = [-2 + 4, 2 + 3] = [2, 5]$$

The geometric construction of the sum **v** + **w** is given in Figure 2.10a (using the triangle law) and in Figure 2.10b (using the parallelogram law).

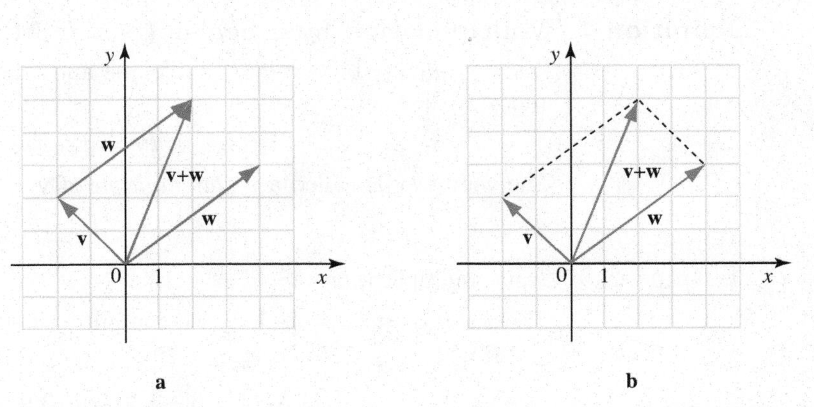

FIGURE 2.10

The sum of the two vectors from Example 2.6

To apply the triangle law, we move one vector (in this case **v**) so that its tail coincides with the head of the other (**w**).

Example 2.7 Total Force Acting on the Tibia

We are ready to answer the question we asked in the introduction to this section. First, we redraw the diagram in Figure 2.1 as shown in Figure 2.11: we choose a point on the tip of the tibia, declare it to be the origin, and draw the coordinate axes as shown. We move all vectors (forces) so that their tails are at the origin. The forces and their magnitudes are given in Table 2.1.

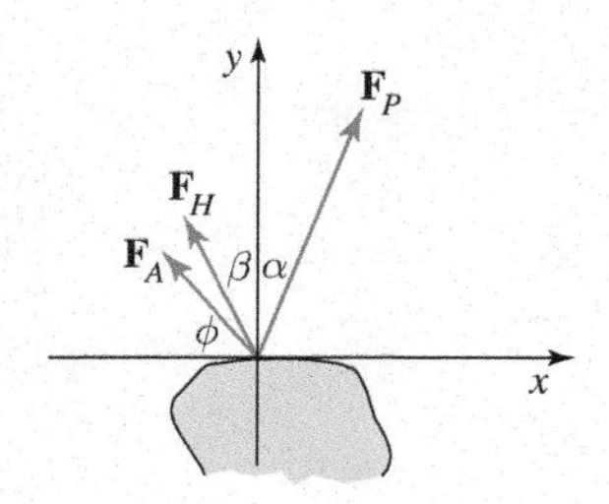

FIGURE 2.11

Forces acting on a tibia

Table 2.1

Force	Magnitude (N)	Angle (°)
$\mathbf{F}_A$ = loading force	770	$\phi = 55$
$\mathbf{F}_P$ = patellar tendon force	2738	$\alpha = 19$
$\mathbf{F}_H$ = hamstring tendon force	790	$\beta = 20$

Since we know the magnitudes of the forces and the angles they make, we use the polar form of a vector. The magnitude of the loading force $\mathbf{F}_A$ is 770, and the angle the force makes with the positive x-axis is $\theta = 180^\circ - \phi = 125^\circ$. Thus

$$\begin{aligned}\mathbf{F}_A &= [\,\|\mathbf{F}_A\|\cos\theta, \|\mathbf{F}_A\|\sin\theta\,]\\ &= [770\cos 125^\circ, 770\sin 125^\circ] \approx [-441.65, 630.75]\end{aligned}$$

The magnitude of the patellar tendon force $\mathbf{F}_P$ is 2738, and the angle the force makes with the positive x-axis is $\theta = 90^\circ - \alpha = 71^\circ$. Thus

$$\begin{aligned}\mathbf{F}_P &= [\,\|\mathbf{F}_P\|\cos\theta, \|\mathbf{F}_P\|\sin\theta\,]\\ &= [2738\cos 71^\circ, 2738\sin 71^\circ] \approx [891.41, 2588.83]\end{aligned}$$

For the hamstring tendon force, $\|\mathbf{F}_H\| = 790$ and $\theta = 90^\circ + \beta = 110^\circ$. Its polar form is

$$\begin{aligned}\mathbf{F}_H &= [\,\|\mathbf{F}_H\|\cos\theta, \|\mathbf{F}_H\|\sin\theta\,]\\ &= [790\cos 110^\circ, 790\sin 110^\circ] \approx [-270.20, 742.36]\end{aligned}$$

The total force acting on the tibia is the sum

$$\begin{aligned}\mathbf{F}_A + \mathbf{F}_P + \mathbf{F}_H &\approx [-441.65, 630.75] + [891.41, 2588.83] + [-270.20, 742.36]\\ &\approx [179.56, 3961.94]\end{aligned}$$

Its magnitude is

$$\|\mathbf{F}_A + \mathbf{F}_P + \mathbf{F}_H\| \approx \sqrt{(179.56)^2 + (3961.94)^2} \approx 3966.01 \text{ N}$$

Definition 7 Multiplication by a Scalar (Scalar Multiplication)

Let $\mathbf{v} = [v_1, v_2]$ be a vector in a plane and t be a real number. The *product* of t and $\mathbf{v}$ is the vector $t\mathbf{v}$ given by

$$t\mathbf{v} = [tv_1, tv_2] \tag{2.4}$$

The vector $t\mathbf{v}$ is called a *scalar multiple* of $\mathbf{v}$.

Example 2.8 Scalar Multiplication

Given the vector $\mathbf{v} = [-3, 2]$, we compute

$$\begin{aligned}3\mathbf{v} &= 3[-3, 2] = [3(-3), 3(2)] = [-9, 6]\\ 1.3\mathbf{v} &= 1.3[-3, 2] = [-3.9, 2.6]\\ -2\mathbf{v} &= (-2)[-3, 2] = [6, -4]\end{aligned}$$

Clearly, $0\mathbf{v} = [0, 0] = \mathbf{0}$. In Figure 2.12 we drew the vector $\mathbf{v} = [-3, 2]$ and the three scalar multiples $3\mathbf{v}$, $1.3\mathbf{v}$, and $-2\mathbf{v}$.

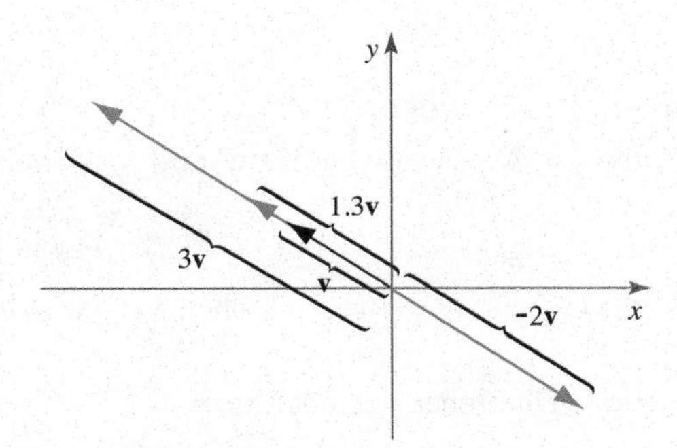

FIGURE 2.12

A vector and its scalar multiples

Note that the vector $t\mathbf{v}$ points in the same direction as $\mathbf{v}$ if $t > 0$ and in the opposite direction if $t < 0$. The length of $t\mathbf{v} = [tv_1, tv_2]$ is

$$\begin{aligned} \|t\mathbf{v}\| &= \sqrt{(tv_1)^2 + (tv_1)^2} \\ &= \sqrt{t^2(v_1^2 + v_2^2)} \\ &= |t|\sqrt{v_1^2 + v_2^2} \\ &= |t|\,\|\mathbf{v}\| \end{aligned} \tag{2.5}$$

In words: the length of $t\mathbf{v}$ is the product of the absolute value of t and the length of $\mathbf{v}$. (The absolute value guarantees that $\|t\mathbf{v}\| \geq 0$.) With this formula in mind, we say that scalar multiplication *scales* a vector: making it shorter if $|t| < 1$ and longer if $|t| > 1$.

Definition 8 Parallel Vectors

Two non-zero vectors are *parallel* if they are scalar multiples of each other. ▲

The vectors $\mathbf{v}$ and $t\mathbf{v}$ (with $t \neq 0$) are parallel vectors. The vectors $\mathbf{v}$, $3\mathbf{v}$, $1.3\mathbf{v}$, and $-2\mathbf{v}$ from Example 2.8 are all parallel to each other.

Assume that $\|\mathbf{v}\| \neq 0$, and let

$$\mathbf{w} = \frac{1}{\|\mathbf{v}\|}\mathbf{v} = \frac{\mathbf{v}}{\|\mathbf{v}\|}$$

The vector $\mathbf{w}$ is a scalar multiple of $\mathbf{v}$ (where the scalar is $t = 1/\|\mathbf{v}\|$), and so $\mathbf{w}$ and $\mathbf{v}$ are parallel. Using (2.5) with $t = 1/\|\mathbf{v}\|$, we get

$$\|\mathbf{w}\| = \left\|\frac{1}{\|\mathbf{v}\|}\mathbf{v}\right\| = \left|\frac{1}{\|\mathbf{v}\|}\right| \|\mathbf{v}\| = \frac{1}{\|\mathbf{v}\|}\|\mathbf{v}\| = 1$$

i.e., $\mathbf{w}$ is a unit vector. We say that $\mathbf{w}$ is the *unit vector in the direction of the vector* $\mathbf{v}$. Alternatively, we say that by computing $\mathbf{w}$, we have *normalized the vector* $\mathbf{v}$.

Example 2.9 Normalizing Vectors

If $\mathbf{v} = [-3, 2]$, then $\|\mathbf{v}\| = \sqrt{13}$; the unit vector in the direction of $\mathbf{v}$ is the vector

$$\mathbf{w} = \frac{1}{\|\mathbf{v}\|}\mathbf{v} = \frac{1}{\sqrt{13}}[-3, 2] = \left[-\frac{3}{\sqrt{13}}, \frac{2}{\sqrt{13}}\right]$$

The unit vector in the direction of $\mathbf{v} = [1, 1]$ is

$$\mathbf{w} = \frac{1}{\|\mathbf{v}\|}\mathbf{v} = \frac{1}{\sqrt{2}}[1, 1] = \left[\frac{1}{\sqrt{2}}, \frac{1}{\sqrt{2}}\right]$$

▲

If $t = -1$, then $t\mathbf{v} = (-1)\mathbf{v}$ is the vector that is of the same length as $\mathbf{v}$ but points in the opposite direction. This vector is denoted by $-\mathbf{v}$ and is called the *negative of the vector* $\mathbf{v}$. So if $\mathbf{v} = [v_1, v_2]$, then $-\mathbf{v} = [-v_1, -v_2]$.

Definition 9 Difference of Vectors

The *difference* $\mathbf{w} - \mathbf{v}$ of vectors $\mathbf{w}$ and $\mathbf{v}$ is the sum of $\mathbf{w}$ and the negative of $\mathbf{v}$:

$$\mathbf{w} - \mathbf{v} = \mathbf{w} + (-\mathbf{v})$$

If $\mathbf{v} = [v_1, v_2]$ and $\mathbf{w} = [w_1, w_2]$, then $\mathbf{w} - \mathbf{v} = [w_1 - v_1, w_2 - v_2]$.

Example 2.10 Subtraction (Difference) of Vectors

Given two points $W = (w_1, w_2)$ and $V = (v_1, v_2)$ in a plane, the directed line segment $\overrightarrow{VW}$ is given by

$$\overrightarrow{VW} = [w_1 - v_1, w_2 - v_2]$$

Since the difference of $\mathbf{w} = [w_1, w_2]$ and $\mathbf{v} = [v_1, v_2]$ is equal to

$$\mathbf{w} - \mathbf{v} = [w_1 - v_1, w_2 - v_2]$$

as well, we conclude that $\overrightarrow{VW}$ is a representative directed line segment of the vector $\mathbf{w} - \mathbf{v}$ that starts at V; see Figure 2.13a.

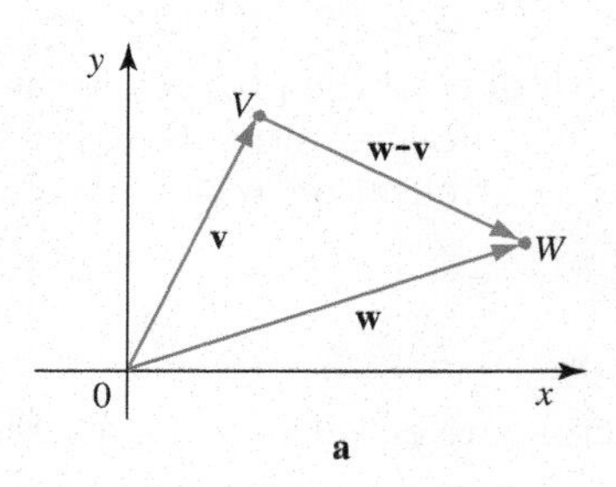

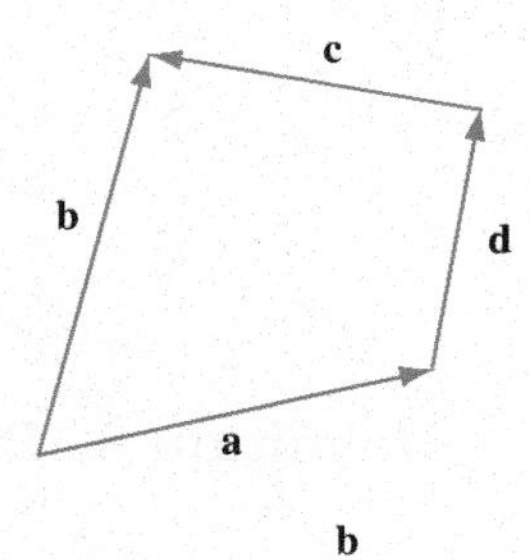

FIGURE 2.13

Finding the difference of vectors

An easy way to remember which way the difference goes is this: the vector (or the directed line segment representing it) $\overrightarrow{VW}$ starts at V and ends at W. An alternative route from V to W is via the origin 0: from V to 0 (which is the vector $-\mathbf{v}$) and then from 0 to W (which is $\mathbf{w}$). Thus,

$$\overrightarrow{VW} = -\mathbf{v} + \mathbf{w} = \mathbf{w} - \mathbf{v}$$

For another illustration of this routine, look at Figure 2.13b: to express the vector $\mathbf{a}$ in terms of vectors $\mathbf{b}$, $\mathbf{c}$, and $\mathbf{d}$, we need to use an alternative route from the tail of $\mathbf{a}$ to its tip. So we walk along $\mathbf{b}$ (so it will be $+\mathbf{b}$), then along $\mathbf{c}$, but in the opposite direction (so the contribution will be $-\mathbf{c}$), and then along $\mathbf{d}$, again in the opposite direction (and the contribution will be $-\mathbf{d}$). Thus,

$$\mathbf{a} = \mathbf{b} - \mathbf{c} - \mathbf{d}$$

In the same way, we find out that $\mathbf{d} = -\mathbf{a} + \mathbf{b} - \mathbf{c}$.

Recall that we defined two special unit vectors, $\mathbf{i} = [1, 0]$ and $\mathbf{j} = [0, 1]$. Using vector addition (subtraction) and multiplication by scalars, we write

$$\begin{aligned}\mathbf{v} = [v_1, v_2] &= [v_1, 0] + [0, v_2] \\ &= v_1[1, 0] + v_2[0, 1] \\ &= v_1\mathbf{i} + v_2\mathbf{j} \qquad (2.6)\end{aligned}$$

Thus, we have *expressed the vector* $\mathbf{v}$ *in terms of the vectors* $\mathbf{i}$ *and* $\mathbf{j}$. For example,

$$\begin{aligned}[4, -2] &= 4\mathbf{i} - 2\mathbf{j} \\ [7, 0] &= 7\mathbf{i} \\ [0, -6] &= -6\mathbf{j}\end{aligned}$$

and so on.

Vectors in Space

All definitions, concepts, and formulas we have introduced extend in a straightforward way to three (or more) dimensions.

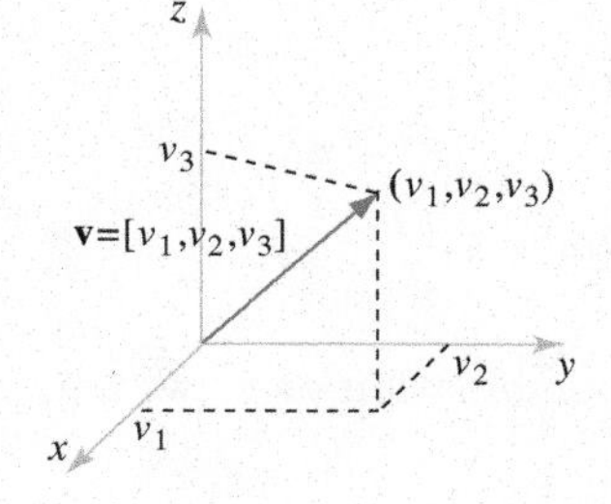

FIGURE 2.14

Correspondence between a vector and a point in space

A vector $\mathbf{v}$ in space is an ordered triple $[v_1, v_2, v_3]$ of real numbers. The set $\mathbb{R}^3$ can be viewed either as a set of points or as a set of vectors (the correspondence is established by declaring the point (v_1, v_2, v_3) to be the head of the vector $\mathbf{v} = [v_1, v_2, v_3]$; see Figure 2.14).

The length of a vector $\mathbf{v} = (v_1, v_2, v_3)$ is given by

$$\|\mathbf{v}\| = \sqrt{v_1^2 + v_2^2 + v_3^2}$$

Addition and subtraction of vectors in $\mathbb{R}^3$, as well as scalar multiplication, are defined component-wise: if $\mathbf{v} = (v_1, v_2, v_3)$ and $\mathbf{w} = (w_1, w_2, w_3)$, then

$$\mathbf{v} \pm \mathbf{w} = [v_1 \pm w_1, v_2 \pm w_2, v_3 \pm w_3]$$

and, for any real number t,

$$t\mathbf{v} = [tv_1, tv_2, tv_3]$$

Example 2.11 **Vector Operations in $\mathbb{R}^3$**

Given $\mathbf{v} = [2, -1, 3]$ and $\mathbf{w} = [3, -1, 0]$, find $3\mathbf{v}$, $2\mathbf{v} - 0.5\mathbf{w}$, $\mathbf{w}/4$, and $\|\mathbf{v}\|$. As well, find the unit vector in the direction of the vector $\mathbf{v}$.

▶ By the definition of scalar multiplication,

$$3\mathbf{v} = 3[2, -1, 3] = [6, -3, 9]$$

Using columns for vectors (for clarity), we write the above calculation as

$$3\mathbf{v} = 3\begin{bmatrix} 2 \\ -1 \\ 3 \end{bmatrix} = \begin{bmatrix} 6 \\ -3 \\ 9 \end{bmatrix}$$

Next,

$$2\mathbf{v} - 0.5\mathbf{w} = 2\begin{bmatrix} 2 \\ -1 \\ 3 \end{bmatrix} - 0.5\begin{bmatrix} 3 \\ -1 \\ 0 \end{bmatrix} = \begin{bmatrix} 4 \\ -2 \\ 6 \end{bmatrix} + \begin{bmatrix} -1.5 \\ 0.5 \\ 0 \end{bmatrix} = \begin{bmatrix} 2.5 \\ -1.5 \\ 6 \end{bmatrix}$$

To calculate $\mathbf{w}/4$, we write

$$\frac{\mathbf{w}}{4} = \frac{1}{4}\mathbf{w} = \frac{1}{4}\begin{bmatrix} 3 \\ -1 \\ 0 \end{bmatrix} = \begin{bmatrix} 3/4 \\ -1/4 \\ 0 \end{bmatrix}$$

The length of $\mathbf{v} = [2, -1, 3]$ is

$$\|\mathbf{v}\| = \sqrt{(2)^2 + (-1)^2 + (3)^2} = \sqrt{14}$$

and the unit vector in the direction of $\mathbf{v}$ is

$$\frac{\mathbf{v}}{\|\mathbf{v}\|} = \frac{1}{\sqrt{14}}\mathbf{v} = \frac{1}{\sqrt{14}}\begin{bmatrix} 2 \\ -1 \\ 3 \end{bmatrix} = \begin{bmatrix} 2/\sqrt{14} \\ -1/\sqrt{14} \\ 3/\sqrt{14} \end{bmatrix}$$ ▲

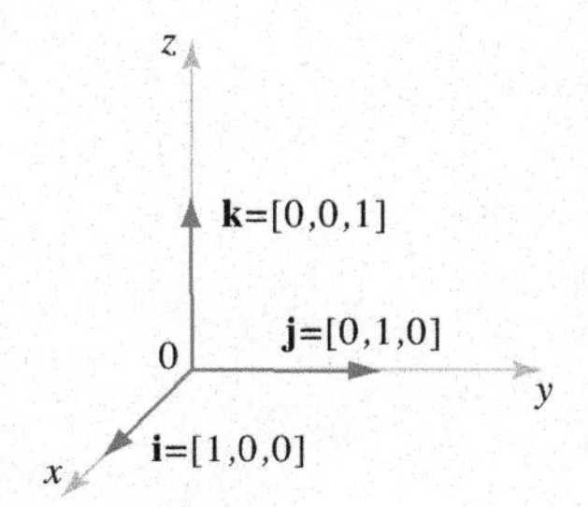

FIGURE 2.15

Unit coordinate vectors in $\mathbb{R}^3$

The three unit (standard) coordinate vectors in $\mathbb{R}^3$ are $\mathbf{i} = [1, 0, 0]$, $\mathbf{j} = [0, 1, 0]$, and $\mathbf{k} = [0, 0, 1]$; see Figure 2.15.

From

$$\begin{aligned} \|\mathbf{v}\| &= [v_1, v_2, v_3] \\ &= v_1[1, 0, 0] + v_2[0, 1, 0] + v_3[0, 0, 1] = v_1\mathbf{i} + v_2\mathbf{j} + v_3\mathbf{k} \end{aligned} \tag{2.7}$$

we conclude that any vector $\mathbf{v}$ in $\mathbb{R}^3$ can be expressed in terms of the unit vectors $\mathbf{i}$, $\mathbf{j}$, and $\mathbf{k}$.

Next, we formalize this idea of *expressing a vector in terms of vectors.*

Definition 10 Linear Combination

A *linear combination* of vectors $\mathbf{v}_1, \mathbf{v}_2, \ldots, \mathbf{v}_k$ is the vector

$$c_1\mathbf{v}_1 + c_2\mathbf{v}_2 + \cdots + c_k\mathbf{v}_k,$$

where $k \geq 1$ and the *coefficients* $c_1, c_2, \ldots, c_k$ are real numbers.

For example, the expression $5\mathbf{v} - 2\mathbf{w} + 0.4\mathbf{z}$ is a linear combination of $\mathbf{v}$, $\mathbf{w}$, and $\mathbf{z}$. The following are linear combinations:

$$1.2\mathbf{v}, \quad \mathbf{w} - 0.3\mathbf{v} - \mathbf{z}, \quad \mathbf{v}/7 - \mathbf{j}, \quad \|\mathbf{v}\|\mathbf{v} + 4\mathbf{w}$$

The vector $[1, -3, -3]$ is a linear combination of the vectors

$$\begin{bmatrix} 1 \\ -1 \\ 0 \end{bmatrix}, \quad \begin{bmatrix} 0 \\ 2 \\ -1 \end{bmatrix}, \quad \text{and} \quad \begin{bmatrix} 1 \\ 1 \\ 1 \end{bmatrix}$$

because

$$\begin{bmatrix} 1 \\ -3 \\ -3 \end{bmatrix} = 3\begin{bmatrix} 1 \\ -1 \\ 0 \end{bmatrix} + 1\begin{bmatrix} 0 \\ 2 \\ -1 \end{bmatrix} - 2\begin{bmatrix} 1 \\ 1 \\ 1 \end{bmatrix}$$

When we write

$$\mathbf{v} = v_1\mathbf{i} + v_2\mathbf{j} + v_3\mathbf{k}$$

we say that we have *expressed the vector* $\mathbf{v}$ *as a linear combination of vectors* $\mathbf{i}$, $\mathbf{j}$, *and* $\mathbf{k}$. (Note that the coefficients in the linear combination are the components of $\mathbf{v}$.)

Properties of Vector Operations

We list the properties of the vector operations that we defined in this section. All formulas can be verified by writing the vectors involved in terms of their coordinates (see Exercises 37 and 38).

Theorem 1 Properties of Vector Operations

Assume that $\mathbf{v}$, $\mathbf{w}$, and $\mathbf{z}$ are vectors in $\mathbb{R}^2$ or in $\mathbb{R}^3$ (or in $\mathbb{R}^n$, $n \geq 2$), and let t and s denote real numbers. Then

(a) *commutativity* of vector addition: $\mathbf{v} + \mathbf{w} = \mathbf{w} + \mathbf{v}$

(b) *associativity* of vector addition: $(\mathbf{v} + \mathbf{w}) + \mathbf{z} = \mathbf{v} + (\mathbf{w} + \mathbf{z})$

(c) zero vector: $\mathbf{v} + \mathbf{0} = \mathbf{v}$

(d) negative of a vector: $\mathbf{v} + (-\mathbf{v}) = \mathbf{0}$

(e) *distributivity* with respect to vector addition: $t(\mathbf{v} + \mathbf{w}) = t\mathbf{v} + t\mathbf{w}$

(f) *distributivity* with respect to scalar multiplication: $(t + s)\mathbf{v} = t\mathbf{v} + s\mathbf{v}$

(g) scalar multiplication: $t(s\mathbf{v}) = (ts)\mathbf{v}$

(h) scalar multiplication by 0 and 1: $0\mathbf{v} = \mathbf{0}$, $1\mathbf{v} = \mathbf{v}$

When we say "vectors in $\mathbb{R}^2$ or in $\mathbb{R}^3$" we mean that *all* vectors involved are in $\mathbb{R}^2$ or *all* vectors involved are in $\mathbb{R}^3$. It does not make any sense to add a two-dimensional vector to a three-dimensional vector.

We will encounter numerous situations where we will need the properties listed in Theorem 1. Perhaps the best way to remember these formulas (what am I allowed to do with vectors?) is to realize that the addition of vectors and the multiplication of a vector by a scalar share the properties of the addition and multiplication of real numbers.

Property (b) states that the order in which three vectors are added is not relevant: we could add the sum of the first and second vectors to the third vector (left side) or add the first vector to the sum of the second and the third vectors (right side). Together with the commutativity in (a), this means that we can add vectors in any order we like. Thus, we do need to use brackets when we write $\mathbf{v} + \mathbf{w} + \mathbf{z}$. (Note that we have already used this fact in Example 2.7, where we were adding three forces.)

Vectors in Applications

We have already mentioned a few applications where vectors are used, and we will encounter more as we start going deeper into various topics in linear algebra. The strength of vectors lies in their ability to describe important physical quantities such as forces and motions. Linear and angular momenta, torque, impulse, position, velocity, and acceleration are all vectors. To study these (and other) quantities, we often use *vector fields,* namely functions whose range is a subset of the set of vectors in $\mathbb{R}^2$ or $\mathbb{R}^3$ (or in higher dimensions). A vector field allows us to assign vectors to points in a plane or in space, depending on the phenomenon that we are studying.

The output of a mathematical model of intra-cellular forces is given in Figure 2.16. Researchers hope that the model will help them understand cell migration patterns. [The illustration has been modelled on the diagrams posted at www.cellmigration.org/resource/modeling/model_approaches.shtml.]

FIGURE 2.16

Vectors depicting intra-cellular forces

Three-dimensional vectors have been used in the study of joint actuation in biological neuromuscular systems. [Source: Krishnaswamy, P., Brown E.N., & Herr, H.M. (2011). Human leg model predicts ankle muscle-tendon morphology, state, roles and energetics in walking. *PLoS Computational Biology,* 7(3): e1001107. doi:10.1371/journal.pcbi.1001107.] The results of the study are important in biomechanics, neuroscience, and prosthetics research. Vectors have been used to gain insight into the organization of a muscle on various size scales. [See Grosberg A., Kuo P.-L., Guo C.-L., Geisse N.A., Bray M.-A., et al. (2011). Self-organization of muscle cell structure and function. *PLoS Computational Biology,* 7(2): e1001088. doi:10.1371/journal.pcbi.1001088.]

Vector fields are an ideal tool for describing all kinds of flows. Figure 2.17a models the motion of sea ice after it has separated from a glacier. Circular flow near a sink can be modelled using a vector field such as the one drawn in Figure 2.17b.

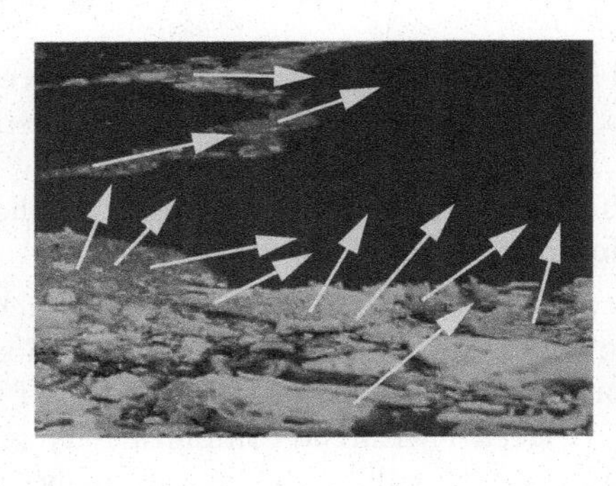

a

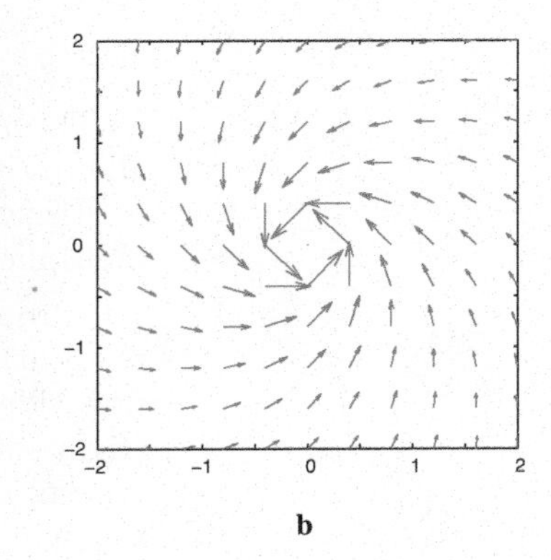

b

FIGURE 2.17
Vector fields
(Courtesy of Miroslav Lovrić.)

Four- and five-dimensional vectors (whose components are not related to length, i.e., are not spatial components) have been used in DNA analysis and the study of large genomes; see Dewey C.N., Huggins P.M., Woods K., Sturmfels B., & Pachter L. (2006). Parametric alignment of drosophila genomes. *PLoS Computational Biology,* 2(6): e73. doi:10.1371/journal.pcbi.0020073.

Summary Quantities such as **force** and **motion,** which are characterized by both **magnitude** and **direction,** are examples of **vectors.** Algebraically, a **vector** is an ordered n-tuple (an ordered pair or an ordered triple, in low dimensions) of real numbers. Geometrically, a vector materializes through a **directed line segment.** Vectors can be **added, subtracted,** and **multiplied by scalars.** As an example of the use of vector operations we defined the **linear combination** of vectors. Every vector in a plane or in space can be expressed as a linear combination of **standard unit vectors.** At the end of the section we mentioned several research papers that use vectors to study a wide range of phenomena in biology and health sciences.

2 Exercises

1. Let $\mathbf{v} = [3, -1]$. Find the representative directed line segments for $\mathbf{v}$ that start at the points $(10, 3)$ and $(-4, -7)$.

2. Let $\mathbf{v} = [-2, 4, -11]$. Find the representative directed line segments for $\mathbf{v}$ that start at the points $(0, 1, 4)$ and $(-2, 0, 0)$.

3. A vector $\mathbf{v}$ is represented by the directed line segment $\overrightarrow{AB}$, where $A = (-4, -1)$ and $B = (3, -2)$. Find the representative directed line segment of $\mathbf{v}$ that starts at $(0, 1)$.

4. A vector $\mathbf{v}$ is represented by the directed line segment $\overrightarrow{AB}$, where $A = (0, 5, -7)$ and $B = (9, -7, 2)$. Find the representative directed line segment of $\mathbf{v}$ that starts at $(1, 2, 3)$.

5. Find the vector in a plane whose length is 15 and whose direction makes a (counterclockwise) angle of $3\pi/4$ radians with respect to the positive x-axis.

6. Find the vector in a plane whose length is 2 and whose direction makes a (counterclockwise) angle of $2\pi/3$ radians with respect to the positive x-axis.

7. If $\|\mathbf{v}\| = 3$, what is the length of the vector $\mathbf{w} = -4\|\mathbf{v}\|\,\mathbf{v}$?

8. Let $\mathbf{v} = [3, 1]$. Find a vector parallel to $\mathbf{v}$ whose length is 13.

9. Let $\mathbf{v} = [-1, 4, 1]$. Find a vector parallel to $\mathbf{v}$ whose length is 10.

10. Which of the three vectors $\mathbf{i} - \mathbf{j} + \mathbf{k}$, $2\mathbf{i} + \mathbf{k}$, and $3\mathbf{j}$ is the longest?

11–22 ▪ Let $\mathbf{a} = [3, -2]$, $\mathbf{b} = [-1, 1]$, $\mathbf{c} = [-1, 0]$, and $\mathbf{d} = [0, 4]$.

11. Find the linear combination $3\mathbf{a} - \mathbf{b} + \mathbf{j}$.

12. Find the linear combination $-4\mathbf{b} + \mathbf{c} - 0.5\mathbf{d}$.

13. Calculate $\|\mathbf{a} + \mathbf{i}\|$.

14. Find $\|\mathbf{d} - \mathbf{c}\|$.

15. Write the vector $\mathbf{b}$ in polar form.

16. Write the vector $\mathbf{d}$ in polar form.

17. Find the unit vector in the direction of $\mathbf{b}$.

18. Find the unit vector in the direction of $\mathbf{a}$.

19. Express the vector $\mathbf{d}$ as a linear combination of the vectors $\mathbf{b}$ and $\mathbf{c}$; i.e., find real numbers α and β so that $\mathbf{d} = \alpha\mathbf{b} + \beta\mathbf{c}$.

20. Express the vector $\mathbf{c}$ as a linear combination of the vectors $\mathbf{a}$ and $\mathbf{b}$.

21. Show that $\|\mathbf{a} + \mathbf{c}\| \neq \|\mathbf{a}\| + \|\mathbf{c}\|$.

22. Show that $\|\mathbf{b} - \mathbf{d}\| \neq \|\mathbf{b}\| - \|\mathbf{d}\|$.

23. Find x and y so that $[5, y] = x[4, 1]$.

24. Find x and y so that $[x, y] - [2, x - 4y] = [4, 7]$.

25. Solve the vector equation $3(2\mathbf{a} - \mathbf{x} - \mathbf{b}) = \mathbf{b} - 2(\mathbf{a} + 7\mathbf{x})$ for $\mathbf{x}$.

26. Solve the vector equation $4\mathbf{b} - 2(\mathbf{v} - 3\mathbf{a} + \mathbf{j}) = \mathbf{v} - 2(\mathbf{j} + \mathbf{b} - 5\mathbf{v})$ for $\mathbf{v}$.

27–34 ▪ Consider the following vectors in $\mathbb{R}^3$:

$$\mathbf{a} = \begin{bmatrix} 4 \\ 0 \\ -2 \end{bmatrix}, \quad \mathbf{b} = \begin{bmatrix} -1 \\ 6 \\ 0 \end{bmatrix}, \quad \text{and} \quad \mathbf{c} = \begin{bmatrix} 2 \\ 7 \\ -4 \end{bmatrix}$$

27. Find the linear combination $-2\mathbf{a} + 3\mathbf{j} - 4\mathbf{b}$.

28. Find the linear combination $\mathbf{c} - \mathbf{k} + 2\mathbf{b}$.

29. Normalize the vector $\mathbf{b}$.

30. Normalize the vector $\mathbf{c} - \mathbf{a}$.

31. Express the vector $\mathbf{a}$ as a linear combination of the vector $\mathbf{c}$ and the unit vectors $\mathbf{i}$ and $\mathbf{k}$.

32. Express the vector $\mathbf{c}$ as a linear combination of the vectors $\mathbf{a}$ and $\mathbf{b}$ and the unit vector $\mathbf{i}$.

33. Show that $\|\mathbf{a} + \mathbf{b}\| < \|\mathbf{a}\| + \|\mathbf{b}\|$.

34. Show that $\|\mathbf{a} + \mathbf{c}\| < \|\mathbf{a}\| + \|\mathbf{c}\|$.

35. Assume that $\mathbf{v} = [v_1, v_2]$ is a non-zero vector and let $\mathbf{w} = [v_1/\sqrt{v_1^2 + v_2^2}, v_2/\sqrt{v_1^2 + v_2^2}]$. Show that $\mathbf{w}$ is a unit vector.

36. Assume that $\mathbf{v} = [v_1, v_2]$ is a non-zero vector. Explain how to use the expression

$$\mathbf{v} = \|\mathbf{v}\| \frac{1}{\|\mathbf{v}\|}\mathbf{v} = \|\mathbf{v}\| \left[\frac{v_1}{\|\mathbf{v}\|}, \frac{v_2}{\|\mathbf{v}\|}\right]$$

to obtain the polar form of a vector. (Hint: Draw a picture.)

37. By writing all vectors involved in terms of their components, prove properties (d), (e), and (h) from Theorem 1 for vectors in $\mathbb{R}^3$.

38. By writing all vectors involved in terms of their components, prove properties (c), (f), and (g) from Theorem 1 for vectors in $\mathbb{R}^3$.

39. By using the triangle law for the addition of vectors, draw a diagram illustrating property (b) from Theorem 1.

40. By using the triangle law for the addition of vectors, draw a diagram illustrating property (e) from Theorem 1 with $t = 2$.

3 The Dot Product

In this section, we introduce a useful way of multiplying vectors called the **dot product.** We will relate it to calculating distances (magnitude of a vector) and angles between vectors. As well, we will use the dot product to obtain equations of lines and planes in Section 4.

The Dot Product

The definition of the dot product is technical and not revealing. However, as we keep working with it, we'll start appreciating its relevance and importance.

Definition 11 The Dot Product

The *dot product* of two vectors $\mathbf{v} = [v_1, v_2]$ and $\mathbf{w} = [w_1, w_2]$ in $\mathbb{R}^2$ is the real number $\mathbf{v} \cdot \mathbf{w}$ defined by

$$\mathbf{v} \cdot \mathbf{w} = v_1 w_1 + v_2 w_2$$

If $\mathbf{v} = [v_1, v_2, v_3]$ and $\mathbf{w} = [w_1, w_2, w_3]$ are vectors in $\mathbb{R}^3$, then

$$\mathbf{v} \cdot \mathbf{w} = v_1 w_1 + v_2 w_2 + v_3 w_3$$

In the same way, we could define the dot product of vectors in $\mathbb{R}^n$ for any dimension $n = 2, 3, 4, \ldots$.

We repeat that the dot product of two vectors is a *real number* (that's why it is also called the *scalar product*). It is computed by adding up the products of the corresponding coordinates of the two vectors. For example, the dot product of $\mathbf{v} = [-2, 1]$ and $\mathbf{w} = [3, 4]$ is

$$\mathbf{v} \cdot \mathbf{w} = \begin{bmatrix} -2 \\ 1 \end{bmatrix} \cdot \begin{bmatrix} 3 \\ 4 \end{bmatrix} = (-2)(3) + (1)(4) = -2$$

Likewise, if $\mathbf{v} = [2, 0, -5]$ and $\mathbf{w} = [0, -6, -6]$, then

$$\mathbf{v} \cdot \mathbf{w} = \begin{bmatrix} 2 \\ 0 \\ -5 \end{bmatrix} \cdot \begin{bmatrix} 0 \\ -6 \\ -6 \end{bmatrix} = (2)(0) + (0)(-6) + (-5)(-6) = 30$$

Example 3.1 Dot Products of Unit Coordinate Vectors

Compute the dot products of the unit coordinate vectors in $\mathbb{R}^2$ and $\mathbb{R}^3$.

▶ In $\mathbb{R}^2$, $\mathbf{i} = [1, 0]$ and $\mathbf{j} = [0, 1]$. Using Definition 11,

$$\mathbf{i} \cdot \mathbf{i} = (1)(1) + (0)(0) = 1$$

Likewise, $\mathbf{j} \cdot \mathbf{j} = 0$. As well,

$$\mathbf{i} \cdot \mathbf{j} = (1)(0) + (0)(1) = 0$$

and $\mathbf{j} \cdot \mathbf{i} = 0$.

In $\mathbb{R}^3$, the unit coordinate vectors are $\mathbf{i} = [1, 0, 0]$, $\mathbf{j} = [0, 1, 0]$, and $\mathbf{k} = [0, 0, 1]$. From Definition 11, we get

$$\mathbf{i} \cdot \mathbf{i} = (1)(1) + (0)(0) + (0)(0) = 1$$

In the same way we compute

$$\mathbf{j} \cdot \mathbf{j} = 1 \quad \text{and} \quad \mathbf{k} \cdot \mathbf{k} = 1$$

and

$$\mathbf{i} \cdot \mathbf{j} = \mathbf{j} \cdot \mathbf{i} = 0, \quad \mathbf{i} \cdot \mathbf{k} = \mathbf{k} \cdot \mathbf{i} = 0, \quad \text{and} \quad \mathbf{j} \cdot \mathbf{k} = \mathbf{k} \cdot \mathbf{j} = 0$$

If $\mathbf{v} = [v_1, v_2]$ is a vector in $\mathbb{R}^2$, then

$$\mathbf{v} \cdot \mathbf{v} = v_1 v_1 + v_2 v_2 = v_1^2 + v_2^2 = \|\mathbf{v}\|^2$$

(In the same way, we check that this formula holds in $\mathbb{R}^3$ as well.) Thus, the length of a vector (*geometric* feature) is related to the dot product (*algebraic* feature) by

$$\|\mathbf{v}\| = \sqrt{\mathbf{v} \cdot \mathbf{v}} \tag{3.1}$$

Example 3.2 Calculating the Distance between Points Using the Dot Product

Compute the distance between the points $A = (3, 2, -7)$ and $B = (0, 4, -2)$ in $\mathbb{R}^3$.

▶ The vector from A to B is $\mathbf{v} = [-3, 2, 5]$, and its length is the distance between the two points. From

$$\|\mathbf{v}\|^2 = \mathbf{v} \cdot \mathbf{v} = (-3)(-3) + (2)(2) + (5)(5) = 38$$

we conclude that the distance between A and B is $\sqrt{38}$. ▲

In order to work with the dot product, we need to establish its properties.

Theorem 2 Properties of the Dot Product

Assume that $\mathbf{v}$, $\mathbf{w}$, and $\mathbf{z}$ are vectors in $\mathbb{R}^2$ or in $\mathbb{R}^3$. Then

(a) the dot product is *commutative:* $\mathbf{v} \cdot \mathbf{w} = \mathbf{w} \cdot \mathbf{v}$

(b) the dot product is *distributive* with respect to vector addition: $\mathbf{v} \cdot (\mathbf{w} + \mathbf{z}) = \mathbf{v} \cdot \mathbf{w} + \mathbf{v} \cdot \mathbf{z}$

(c) $(t\mathbf{v}) \cdot \mathbf{w} = t(\mathbf{v} \cdot \mathbf{w})$ for any real number t ▲

Note that all equal signs in the theorem establish equalities between real numbers. Let's look at formula (c) for a moment: the expression $t\mathbf{v}$ is the product of a real number and a vector (i.e., it is a scalar multiple of a vector, and thus a vector); the dot in $(t\mathbf{v}) \cdot \mathbf{w}$ represents the dot product of two vectors. On the right side, the dot in $\mathbf{v} \cdot \mathbf{w}$ is the dot product of two vectors (hence a real number); finally, the product $t(\mathbf{v} \cdot \mathbf{w})$ is the product of two real numbers. (So in this formula we have three different multiplications.)

All three formulas in Theorem 2 can be proven by writing the vectors involved in terms of their coordinates and using appropriate definitions (see Exercise 25).

Example 3.3 Calculating the Dot Product Using Theorem 2

Instead of using square brackets, we can write vectors as linear combinations of the unit coordinate vectors. In that case, we use properties (a) to (c) of Theorem 2 to calculate the dot product.

Here is an example: let $\mathbf{v} = 4\mathbf{i} - 2\mathbf{j}$ and $\mathbf{w} = 3\mathbf{i} - \mathbf{j}$. Then

$$\begin{aligned}
\mathbf{v} \cdot \mathbf{w} &= (4\mathbf{i} - 2\mathbf{j}) \cdot (3\mathbf{i} - \mathbf{j}) \\
&= 4\mathbf{i} \cdot (3\mathbf{i} - \mathbf{j}) - 2\mathbf{j} \cdot (3\mathbf{i} - \mathbf{j}) \\
&= 4(\mathbf{i} \cdot (3\mathbf{i} - \mathbf{j})) - 2(\mathbf{j} \cdot (3\mathbf{i} - \mathbf{j})) \\
&= 4(3\mathbf{i} \cdot \mathbf{i} - \mathbf{i} \cdot \mathbf{j}) - 2(3\mathbf{j} \cdot \mathbf{i} - \mathbf{j} \cdot \mathbf{j}) \\
&= 4(3 - 0) - 2(0 - 1) = 14
\end{aligned}$$

using the dot products we calculated in Example 3.1. ▲

To make a long story short, Theorem 2 tells us that we can calculate the dot product in the same way as we calculate the product of two binomials (or trinomials): by multiplying each term of the first factor by each term of the second factor.

Our definition of the dot product is algebraic: it tells us how to compute $\mathbf{v} \cdot \mathbf{w}$ when the coordinates of $\mathbf{v}$ and $\mathbf{w}$ are known. But vectors are geometric objects as well—how do we find the dot product of vectors if we know their magnitudes and directions?

First, we need to define the angle between vectors.

Definition 12 Angle between Vectors

Assume that $\mathbf{v}$ and $\mathbf{w}$ are non-zero vectors, placed so that their tails are located at the same point. The *angle θ between* $\mathbf{v}$ *and* $\mathbf{w}$ is the smaller of the two angles formed by the directions of $\mathbf{v}$ and $\mathbf{w}$. If the two angles are equal, then $\theta = \pi$ (radians).

Thus, the angle between two vectors satisfies $0 \leq \theta \leq \pi$; see Figure 3.1.

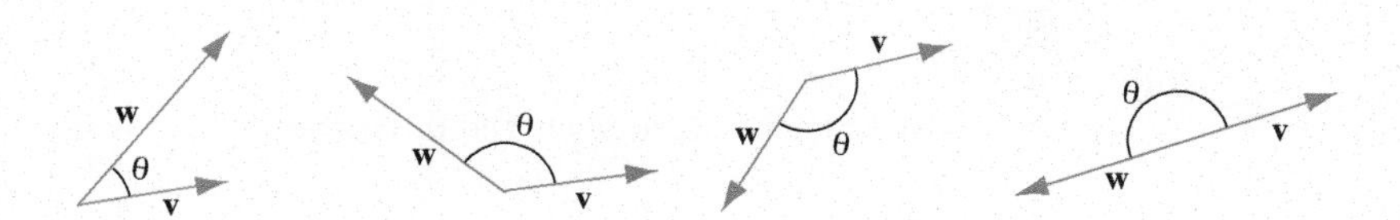

FIGURE 3.1
Angle between vectors

Consider the triangle formed by two non-zero vectors $\mathbf{v}$ and $\mathbf{w}$ and their difference $\mathbf{w} - \mathbf{v}$; see Figure 3.2a. In the calculation that follows we will use the fact that $\|\mathbf{a}\|^2 = \mathbf{a} \cdot \mathbf{a}$ for any vector $\mathbf{a}$.

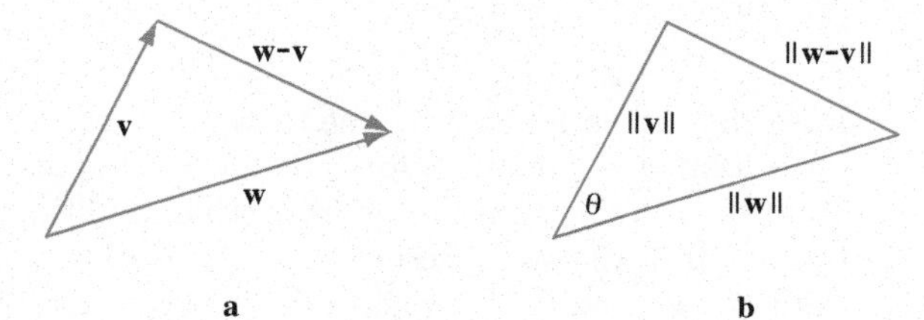

FIGURE 3.2
Calculating $\|\mathbf{w} - \mathbf{v}\|$

First, redraw the triangle in Figure 3.2a and label the lengths, rather than the vectors (Figure 3.2b). Applying the law of cosines, we get

$$\|\mathbf{w} - \mathbf{v}\|^2 = \|\mathbf{w}\|^2 + \|\mathbf{v}\|^2 - 2\|\mathbf{w}\| \, \|\mathbf{v}\| \cos\theta \tag{3.2}$$

where θ is the angle between $\mathbf{w}$ and $\mathbf{v}$. Expanding $\|\mathbf{w}-\mathbf{v}\|^2$ using $\|\mathbf{a}\|^2 = \mathbf{a}\cdot\mathbf{a}$ with $\mathbf{a} = \mathbf{w} - \mathbf{v}$, and with the help of the properties of the dot product from Theorem 2, we obtain

$$\begin{aligned}
\|\mathbf{w} - \mathbf{v}\|^2 &= (\mathbf{w} - \mathbf{v}) \cdot (\mathbf{w} - \mathbf{v}) \\
&= \mathbf{w} \cdot \mathbf{w} - \mathbf{v} \cdot \mathbf{w} - \mathbf{w} \cdot \mathbf{v} + \mathbf{v} \cdot \mathbf{v} \\
&= \|\mathbf{w}\|^2 - \mathbf{v} \cdot \mathbf{w} - \mathbf{v} \cdot \mathbf{w} + \|\mathbf{v}\|^2 \\
&= \|\mathbf{w}\|^2 - 2\mathbf{v} \cdot \mathbf{w} + \|\mathbf{v}\|^2
\end{aligned} \tag{3.3}$$

Note that we used $\|\mathbf{a}\|^2 = \mathbf{a}\cdot\mathbf{a}$ two more times (going from the second to the third line). As well, we used the commutativity of the dot product ($\mathbf{v} \cdot \mathbf{w} = \mathbf{w} \cdot \mathbf{v}$).

Combining (3.2) and (3.3) and simplifying, we obtain

$$\begin{aligned}
\|\mathbf{w}\|^2 - 2\mathbf{v} \cdot \mathbf{w} + \|\mathbf{v}\|^2 &= \|\mathbf{w}\|^2 + \|\mathbf{v}\|^2 - 2\|\mathbf{w}\| \, \|\mathbf{v}\| \cos\theta \\
-2\mathbf{v} \cdot \mathbf{w} &= -2\|\mathbf{w}\| \, \|\mathbf{v}\| \cos\theta
\end{aligned}$$

and therefore

$$\mathbf{v} \cdot \mathbf{w} = \|\mathbf{v}\| \, \|\mathbf{w}\| \cos\theta \tag{3.4}$$

So, to calculate the dot product of two vectors, we multiply the product of their lengths by the cosine of the angle between them.

Example 3.4 Calculating the Dot Product

Suppose that, for the vectors in Figure 3.2, $\|\mathbf{v}\| = 5$, $\|\mathbf{w}\| = 7$, and $\theta = \pi/4$. Then

$$\mathbf{v} \cdot \mathbf{w} = \|\mathbf{v}\| \, \|\mathbf{w}\| \cos(\pi/4) = 5 \cdot 7 \cdot \frac{\sqrt{2}}{2} \approx 24.75$$

From (3.4) we obtain the formula

$$\cos\theta = \frac{\mathbf{v} \cdot \mathbf{w}}{\|\mathbf{v}\| \, \|\mathbf{w}\|} \tag{3.5}$$

for the angle between two non-zero vectors $\mathbf{v}$ and $\mathbf{w}$.

Example 3.5 Angle between Vectors in $\mathbb{R}^2$

Find the angle between the vectors $\mathbf{v} = [-2, 1]$ and $\mathbf{w} = [1, 6]$.

From $\mathbf{v} \cdot \mathbf{w} = -2 + 6 = 4$, $\|\mathbf{v}\| = \sqrt{5}$, and $\|\mathbf{w}\| = \sqrt{37}$, we get

$$\cos\theta = \frac{\mathbf{v} \cdot \mathbf{w}}{\|\mathbf{v}\| \, \|\mathbf{w}\|} = \frac{4}{\sqrt{5}\sqrt{37}} \approx 0.294$$

and so $\theta \approx 1.272$ (radians).

Example 3.6 Angle between Vectors in $\mathbb{R}^3$

Find the angle between the vector $\mathbf{v} = [1, 4, 6]$ and the z-axis.

Since the vector $\mathbf{k}$ points in the direction of the z-axis, we need to calculate the angle between $\mathbf{v}$ and $\mathbf{k}$. From

$$\mathbf{v} \cdot \mathbf{k} = (1)(0) + (4)(0) + (6)(1) = 6$$
$$\|\mathbf{v}\| = \sqrt{1 + 16 + 36} = \sqrt{53}$$
$$\|\mathbf{k}\| = \sqrt{1} = 1$$

we compute

$$\cos\theta = \frac{\mathbf{v} \cdot \mathbf{k}}{\|\mathbf{v}\| \, \|\mathbf{k}\|} = \frac{6}{\sqrt{53}} \approx 0.824$$

Thus, the angle between $\mathbf{v}$ and the z-axis is $\theta \approx 0.602$ (radians).

Definition 13 Orthogonal Vectors

Two non-zero vectors are called *orthogonal* (or *perpendicular*) if the angle θ between them is the right angle, i.e., $\theta = \pi/2$ radians.

If $\mathbf{v}$ and $\mathbf{w}$ are orthogonal, then

$$\mathbf{v} \cdot \mathbf{w} = \|\mathbf{v}\| \, \|\mathbf{w}\| \cos(\pi/2) = 0$$

The reverse of this statement is true as well: assume that the dot product of two non-zero vectors $\mathbf{v}$ and $\mathbf{w}$ is zero. Then from

$$\mathbf{v} \cdot \mathbf{w} = \|\mathbf{v}\| \, \|\mathbf{w}\| \cos\theta = 0$$

we conclude that $\cos\theta = 0$ ($\mathbf{v}$ and $\mathbf{w}$ are assumed to be non-zero vectors, so their lengths are positive, not zero); thus, $\theta = \pi/2$, i.e., $\mathbf{v}$ and $\mathbf{w}$ are orthogonal. We have just proven the following statement:

Theorem 3 Orthogonal Vectors and the Dot Product

Two non-zero vectors $\mathbf{v}$ and $\mathbf{w}$ are orthogonal if and only if $\mathbf{v} \cdot \mathbf{w} = 0$.

The theorem tells us that we can use the dot product to determine whether or not two vectors are orthogonal.

Earlier in the section, we showed that the standard unit vectors $\mathbf{i}$ and $\mathbf{j}$ in $\mathbb{R}^2$ and $\mathbf{i}$, $\mathbf{j}$, and $\mathbf{k}$ in $\mathbb{R}^3$ are orthogonal.

Example 3.7 Orthogonal Vectors

Find all non-zero vectors in $\mathbb{R}^2$ that are orthogonal to the vector $\mathbf{v} = [2, -1]$.

Looking at Figure 3.3, we see that there are many vectors that are orthogonal to the given vector $\mathbf{v}$. To identify them all, we look for vectors in the form $\mathbf{w} = [w_1, w_2]$ such that

$$\mathbf{v} \cdot \mathbf{w} = 2w_1 - w_2 = 0$$

We have one equation and two variables, so we declare one of the variables (say, w_2) to be a parameter; we write $w_2 = t$, where t is a real number. From $2w_1 - w_2 = 2w_1 - t = 0$ we get $w_1 = t/2$. Thus, all vectors of the form

$$\mathbf{w} = \begin{bmatrix} t/2 \\ t \end{bmatrix} = t \begin{bmatrix} 1/2 \\ 1 \end{bmatrix}$$

where $t \neq 0$, are orthogonal to $\mathbf{v} = [2, -1]$.

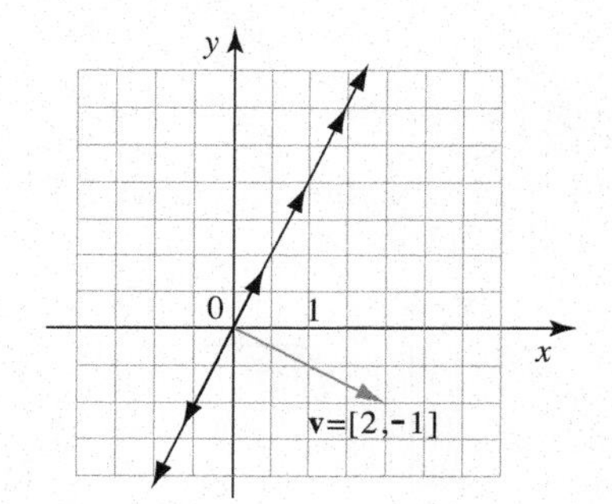

FIGURE 3.3

Vectors orthogonal to $\mathbf{v} = [2, -1]$

Figure 3.3 suggests that all vectors with the required property lie on a line; so, our answer must describe a line. We explore this idea further in the next section.

Summary We can calculate the **dot product** of two vectors in two ways: by multiplying respective components and adding up the products, or by multiplying the product of the lengths of the two vectors by the cosine of the angle between them. Non-zero vectors are **orthogonal** if the angle between them is the right angle. Orthogonal vectors are characterized by the fact that their dot product is zero. The standard unit vectors $\mathbf{i}$ and $\mathbf{j}$ in $\mathbb{R}^2$ and $\mathbf{i}$, $\mathbf{j}$, and $\mathbf{k}$ in $\mathbb{R}^3$ are orthogonal.

3 Exercises

1. What can you say about the angle between two vectors if their dot product is negative?

2. Is the angle between $\mathbf{v} = [8, 2]$ and $\mathbf{w} = [-2, 1]$ acute or obtuse?

3. Assume that $\mathbf{v}$ and $\mathbf{w}$ are vectors in $\mathbb{R}^3$. Explain why the expression $\mathbf{v} \cdot \mathbf{w} + \mathbf{w}$ makes no sense.

4. If $\mathbf{v}$ is a vector, is the expression $\mathbf{v} \cdot (\mathbf{v} \cdot \mathbf{v})$ a real number or a vector?

5. If $\mathbf{v}$ is a vector, is the expression $\|\mathbf{v}\|(\mathbf{v} \cdot \mathbf{v})$ a real number or a vector?

6. Assume that $\mathbf{v}$ and $\mathbf{w}$ are vectors in $\mathbb{R}^3$. Explain what kind of multiplication (between real numbers, between a real number and a vector, or the dot product) is represented by each dot: $5 \cdot (\mathbf{v} \cdot (\mathbf{w} \cdot \mathbf{v}))$. Is the resulting quantity a real number or a vector?

7. Assume that $\mathbf{v}$ and $\mathbf{w}$ are vectors in $\mathbb{R}^3$. Explain what kind of multiplication (of two real numbers, between a real number and a vector, or the dot product) is represented by each dot: $(\mathbf{v} \cdot \mathbf{w}) \cdot (\mathbf{w} \cdot \mathbf{v})$. Is the resulting quantity a real number or a vector?

8–13 ▪ Find the dot product $\mathbf{v} \cdot \mathbf{w}$.

8. $\mathbf{v} = \begin{bmatrix} -9 \\ 1 \end{bmatrix}$, $\mathbf{w} = \begin{bmatrix} 11 \\ -1 \end{bmatrix}$

9. $\mathbf{v} = \begin{bmatrix} -2 \\ 1 \end{bmatrix}$, $\mathbf{w} = \begin{bmatrix} 2 \\ -1 \end{bmatrix}$

10. $\mathbf{v} = \begin{bmatrix} 1 \\ -4 \\ 2 \end{bmatrix}$, $\mathbf{w} = \begin{bmatrix} 0 \\ 11 \\ 2 \end{bmatrix}$

11. $\mathbf{v} = \begin{bmatrix} 4 \\ 8 \\ 0 \end{bmatrix}$, $\mathbf{w} = \begin{bmatrix} 0 \\ 0 \\ 3 \end{bmatrix}$

12. $\mathbf{v} = 3\mathbf{i} - 4\mathbf{j}$, $\mathbf{w} = -4\mathbf{j}$

13. $\mathbf{v} = 3\mathbf{j} + 2\mathbf{k}$, $\mathbf{w} = 5\mathbf{i} + 4\mathbf{j} - 3\mathbf{k}$

14. Describe all of the vectors in a plane that are orthogonal to the vector $\mathbf{v} = [-1, -4]$.

15. Describe all of the vectors in a plane that are orthogonal to the vector $\mathbf{v} = [2, -5]$.

16. Describe all vectors in $\mathbb{R}^3$ which are orthogonal to the vector $\mathbf{v} = [-1, 2, 4]$.

17. Describe all vectors in $\mathbb{R}^3$ which are orthogonal to the vector $\mathbf{v} = [1, -5, 6]$.

18–21 ▪ Find the angle between $\mathbf{v}$ and $\mathbf{w}$.

18. $\mathbf{v} = \begin{bmatrix} -4 \\ 2 \end{bmatrix}$, $\mathbf{w} = \begin{bmatrix} 2 \\ 4 \end{bmatrix}$

19. $\mathbf{v} = \begin{bmatrix} 1 \\ 2 \\ 0 \end{bmatrix}$, $\mathbf{w} = \begin{bmatrix} 0 \\ 3 \\ 1 \end{bmatrix}$

20. $\mathbf{v} = \mathbf{i} + \mathbf{j}$, $\mathbf{w} = 2\mathbf{i}$

21. $\mathbf{v} = 2\mathbf{i} + 4\mathbf{k}$, $\mathbf{w} = -\mathbf{i} - 2\mathbf{j} + 5\mathbf{k}$

22. Find all values of a so that the vectors $\mathbf{v} = [a, 3, 1]$ and $\mathbf{w} = [a, a, 0]$ are orthogonal.

23. Show, in two different ways, that the triangle with vertices at $(1, 1, 5)$, $(3, 2, 7)$, and $(3, -3, 5)$ is a right-angled triangle.

24. Show by example that $\mathbf{u} \cdot \mathbf{v} = \mathbf{u} \cdot \mathbf{w}$ does not imply that $\mathbf{v} = \mathbf{w}$. (Thus, there is no cancellation law for the dot product.)

25. Let $\mathbf{v} = [v_1, v_2]$, $\mathbf{w} = [w_1, w_2]$, and $\mathbf{z} = [z_1, z_2]$ be vectors in $\mathbb{R}^2$.

    (a) Using the definition of the dot product, show that $\mathbf{v} \cdot \mathbf{w} = \mathbf{w} \cdot \mathbf{v}$.

    (b) Using the definition of the dot product, show that the real numbers $\mathbf{v} \cdot (\mathbf{w} + \mathbf{z})$ and $\mathbf{v} \cdot \mathbf{w} + \mathbf{v} \cdot \mathbf{z}$ are equal.

    (c) Use the definition of scalar multiplication and of the dot product to prove that $(t\mathbf{v}) \cdot \mathbf{w} = t(\mathbf{v} \cdot \mathbf{w})$.

26. Take two non-zero vectors $\mathbf{v}$ and $\mathbf{w}$ of the same length, i.e., $\|\mathbf{v}\| = \|\mathbf{w}\|$. Show that the vectors $\mathbf{v} + \mathbf{w}$ and $\mathbf{v} - \mathbf{w}$ are orthogonal.

27. Assume that $\mathbf{v}$ and $\mathbf{w}$ are non-zero vectors. Prove the following:

    (a) If $\mathbf{v}$ and $\mathbf{w}$ are orthogonal, then $\|\mathbf{v} + \mathbf{w}\|^2 = \|\mathbf{v}\|^2 + \|\mathbf{w}\|^2$.

    (b) If $\|\mathbf{v} + \mathbf{w}\|^2 = \|\mathbf{v}\|^2 + \|\mathbf{w}\|^2$, then $\mathbf{v}$ and $\mathbf{w}$ are orthogonal.

28. Show that $(\mathbf{v} + \mathbf{w}) \cdot (\mathbf{v} - \mathbf{w}) = \|\mathbf{v}\|^2 - \|\mathbf{w}\|^2$ holds for any two vectors $\mathbf{v}$ and $\mathbf{w}$.

4 Equations of Lines and Planes

Using our knowledge of vectors and vector operations we now study **lines** in a plane and in space, and **planes** in space. As we will soon see, this will help us reason about, and solve, systems of linear equations.

Lines in a Plane and in Space

There will be a number of situations in this section (and later) where we will use the identification between vectors and points. Recall that to every vector $\mathbf{v} = [v_1, v_2]$ in $\mathbb{R}^2$ we assign the point (v_1, v_2) that represents its head, and vice versa. The vector $\mathbf{v} = [v_1, v_2]$ is called the *position vector* of the point (v_1, v_2). The same identification between points and vectors holds in $\mathbb{R}^3$ as well.

Take a non-zero vector $\mathbf{v} = [v_1, v_2]$ and draw the line l that contains the origin and has the same direction as $\mathbf{v}$; see Figure 4.1. We say that the line l "contains the vector $\mathbf{v}$," or that the vector $\mathbf{v}$ "lies on l." Note that l contains not only $\mathbf{v}$, but also all its scalar multiples $t\mathbf{v}$, where $t \in \mathbb{R}$.

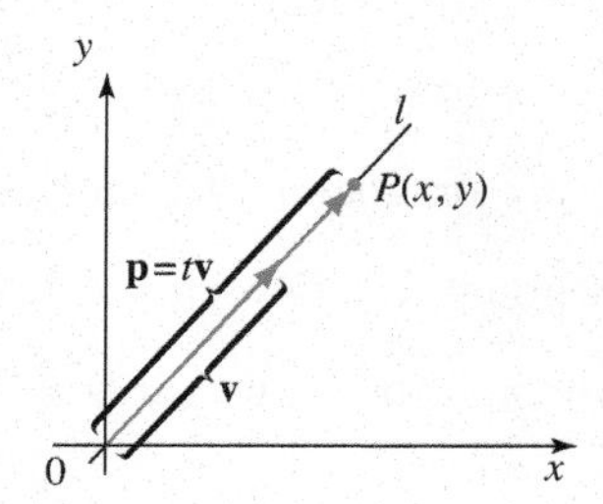

FIGURE 4.1

The line l contains scalar multiples of $\mathbf{v}$

Pick a point $P = (x, y)$ on l, and let $\mathbf{p} = [x, y]$ be the position vector of P. Since $\mathbf{p}$ and $\mathbf{v}$ are parallel, there must be a real number t such that

$$\mathbf{p} = t\mathbf{v} \tag{4.1}$$

(note that t is negative if $\overrightarrow{0P}$ points in the direction opposite to $\mathbf{v}$). In this way, for every point P on l we can find a scalar multiple $t\mathbf{v}$ of $\mathbf{v}$ whose head is at P.

We have just discovered that the line l is "built" of all scalar multiples of $\mathbf{v}$:

$$l = \{t\mathbf{v} \mid t \in \mathbb{R}\} \tag{4.2}$$

The precise meaning of (4.2) is the following: for every point P on l there is a unique value of t for which the head of the vector $t\mathbf{v}$ is at P; as well, each real number t defines a unique point on the line l, given by the head of the vector $t\mathbf{v}$. In other words, there is a one-to-one correspondence

$$\text{points on the line} \leftrightarrow \text{real numbers}$$

Writing (4.1) in coordinates, we obtain

$$\begin{bmatrix} x \\ y \end{bmatrix} = t \begin{bmatrix} v_1 \\ v_2 \end{bmatrix}$$

or

$$\begin{cases} x = tv_1 \\ y = tv_2 \end{cases} \tag{4.3}$$

where the *parameter* t is a real number. The set of equations (4.3) represents the *parametric equation* (or *parametric equations*) of a line through the origin whose direction is given by the vector $\mathbf{v} = [v_1, v_2]$.

We now extend our reasoning to obtain parametric equations of any line (not necessarily going through the origin).

Pick a point $A = (a_1, a_2)$ in $\mathbb{R}^2$ and a vector $\mathbf{v} = [v_1, v_2]$ and construct the line l through A in the direction of $\mathbf{v}$; see Figure 4.2.

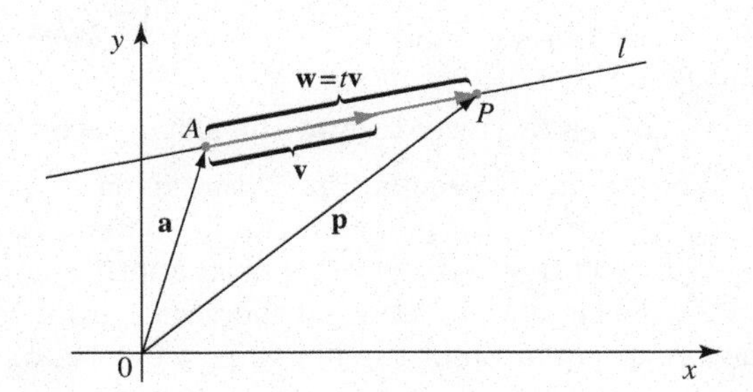

FIGURE 4.2

Obtaining parametric equations of a line

Take any point $P = (x, y)$ on l, and let $\mathbf{p} = [x, y]$ be the vector from 0 to P. Denote by $\mathbf{a} = [a_1, a_2]$ the position vector of the point A. Using the triangle law for the addition of vectors, we write

$$\mathbf{p} = \mathbf{a} + \mathbf{w}$$

where $\mathbf{w}$ is the vector from A to P. Since $\mathbf{w}$ is parallel to $\mathbf{v}$, there is a real number t such that $\mathbf{w} = t\mathbf{v}$, and so

$$\mathbf{p} = \mathbf{a} + t\mathbf{v}$$

This equation is called the *vector form of the parametric equation of a line.* Perhaps the easiest way to remember this is as

$$\text{line} = \text{point} + \text{parameter} \cdot \text{direction vector}$$

In coordinates,

$$\begin{bmatrix} x \\ y \end{bmatrix} = \begin{bmatrix} a_1 \\ a_2 \end{bmatrix} + t \begin{bmatrix} v_1 \\ v_2 \end{bmatrix} \tag{4.4}$$

or

$$\begin{cases} x = a_1 + tv_1 \\ y = a_2 + tv_2 \end{cases} \tag{4.5}$$

where $t \in \mathbb{R}$. Equations (4.5) are referred to as the *parametric equation* (or the *parametric equations*) *of a line.*

In a sequence of examples we work on gaining experience with vector and parametric equations.

Example 4.1 Parametric Equations

(a) Find the parametric equations of the line through the origin that contains the point $(-2, 4)$. What is its slope?

(b) Find the parametric equations of the line of slope $3/2$ that goes through the origin.

▶ (a) The direction vector from 0 to $(-2, 4)$ is $\mathbf{v} = [-2, 4]$. Using (4.4), we write the vector equation of the line as $\mathbf{p} = \mathbf{0} + t\mathbf{v}$, i.e.,

$$\begin{bmatrix} x \\ y \end{bmatrix} = \begin{bmatrix} 0 \\ 0 \end{bmatrix} + t \begin{bmatrix} -2 \\ 4 \end{bmatrix} = t \begin{bmatrix} -2 \\ 4 \end{bmatrix}$$

Writing out the coordinates, we obtain the parametric equations

$$\begin{cases} x = -2t \\ y = 4t \end{cases}$$

where $t \in \mathbb{R}$.

A long way to find the slope is to convert the parametric equations into a point-slope equation. (We do it this way to demonstrate how the conversion is done.) In order to do so, we have to eliminate t; from $x = -2t$ we get $t = -x/2$, which we substitute into $y = 4t$:

$$y = 4t = 4\left(-\frac{x}{2}\right) = -2x$$

Thus, the slope of the line is -2.

What is a fast way to find the slope? See Exercise 3.

(b) The point-slope equation of the given line is $y = 3x/2$. To find parametric equations, we need a point on the line (we actually have it—the origin) and the direction vector $\mathbf{v}$. To get $\mathbf{v}$, we need another point on the line; for instance, when $x = 2$, then $y = 3$. So let $\mathbf{v}$ be the vector from the origin to $(2, 3)$, i.e., $\mathbf{v} = [2, 3]$. The vector equation of the line is $\mathbf{p} = \mathbf{0} + t\mathbf{v}$, or

$$\begin{bmatrix} x \\ y \end{bmatrix} = \begin{bmatrix} 0 \\ 0 \end{bmatrix} + t\begin{bmatrix} 2 \\ 3 \end{bmatrix}$$

and the parametric equations are

$$\begin{cases} x = 2t \\ y = 3t \end{cases} \tag{4.6}$$

where $t \in \mathbb{R}$.

Alternatively, we solve the equation $y = 3x/2$ for x and y. There are two variables, so we declare one to be a parameter: let $x = s$; then $y = 3x/2 = 3s/2$, and thus

$$\begin{bmatrix} x \\ y \end{bmatrix} = \begin{bmatrix} s \\ 3s/2 \end{bmatrix} = s\begin{bmatrix} 1 \\ 3/2 \end{bmatrix}$$

In parametric form,

$$\begin{cases} x = s \\ y = 3s/2 \end{cases} \tag{4.7}$$

where $s \in \mathbb{R}$. Note that the vectors $[2, 3]$ and $[1, 3/2]$ are parallel. Thus, the two parametric equations represent the same line. ▲

Why did we use different symbols for the parameters in (4.6) and (4.7) if both sets of equations represent the same line?

Pick a value for t, say $t = 2$. The corresponding point on the line (substitute $t = 2$ into (4.6)) is $(x = 4, y = 6)$. We do *not* get the same point if we use $s = 2$ in (4.7); we get the point $(x = 2, y = 3)$ instead. However, taking $s = 4$ in (4.7) yields the point $(x = 4, y = 6)$, the same point that we obtained from (4.6) with $t = 2$.

This works the other way around as well: for $s = -2$, equations (4.7) generate the point $(-2, -3)$. We obtain the same point from (4.6) if we use $t = -1$. See Figure 4.3.

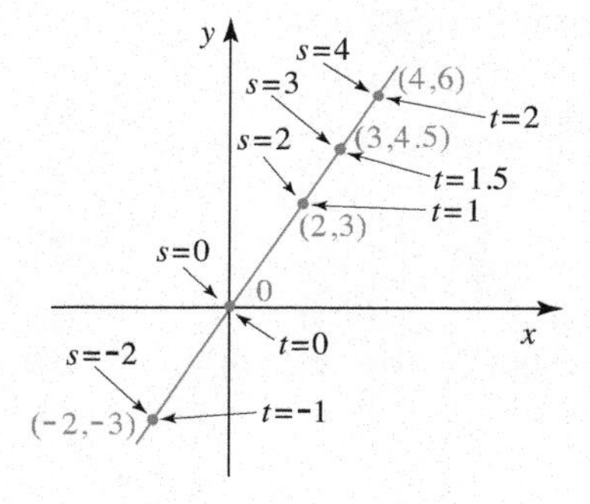

FIGURE 4.3

Two parametrizations: two views of the same line

To summarize: both equations (4.6) and (4.7) do describe the same set of points; however, the same numeric value for s and t does not (in general) yield the same point.

A line has many (infinitely many) different parametric equations (see Exercise 42). Thus, equations that do not *look the same* might represent the same line.

One way to visualize the vector equation

$$\mathbf{p} = \mathbf{a} + t\mathbf{v}$$

is to think of it as describing the motion of a particle along a line, with t representing the time. Thus, each point on the line is identified by the moment (value of t) when the particle reaches it. Sometimes we refer to the point corresponding to $t = 0$ (i.e., the point $\mathbf{a}$) as the *initial point;* negative values of t represent the past. The direction of the vector $\mathbf{v}$ gives the direction of motion, and its magnitude $\|\mathbf{v}\|$ is the constant speed at which the particle moves (see Exercise 18).

This way of looking at vector (and the corresponding parametric) equations is very helpful, for instance, when we study functions of several variables.

Example 4.2 Parametric Equations

Find the parametric equations of the line that goes through the points $A = (-3, 2)$ and $B = (1, 4)$.

▶ For the direction, we take the vector $\mathbf{v}$ from A to B; thus, $\mathbf{v} = [4, 2]$. Using

$$\text{line} = \text{point} + \text{parameter} \cdot \text{direction vector}$$

and taking the *point* to be A, we obtain

$$\begin{bmatrix} x \\ y \end{bmatrix} = \begin{bmatrix} -3 \\ 2 \end{bmatrix} + t \begin{bmatrix} 4 \\ 2 \end{bmatrix}$$

Written in coordinates,

$$\begin{cases} x = -3 + 4t \\ y = 2 + 2t \end{cases}$$

where $t \in \mathbb{R}$.

Let us think a bit about this process. In calculating the parametric equations we made choices: we used the point A (and not B), and we took $\mathbf{v}$ to go from A to B (and not in the reverse direction). What would happen if we recomputed the parametric equations with other possible choices?

First, we convert the equations we obtained into a point-slope equation. To do so, we eliminate t; from $x = -3 + 4t$ we get $t = x/4 + 3/4$, and thus

$$y = 2 + 2t = 2 + 2\left(\frac{x}{4} + \frac{3}{4}\right) = \frac{1}{2}x + \frac{7}{2}$$

Now we modify our choices: using the point $B = (1, 4)$ instead of A, we get

$$\begin{bmatrix} x \\ y \end{bmatrix} = \begin{bmatrix} 1 \\ 4 \end{bmatrix} + s \begin{bmatrix} 4 \\ 2 \end{bmatrix}$$

where $s \in \mathbb{R}$ (we have already used t, so we need a new symbol for the parameter); the corresponding parametric equations are

$$\begin{cases} x = 1 + 4s \\ y = 4 + 2s \end{cases}$$

$s \in \mathbb{R}$. These equations are not the same as those we got the first time. Nevertheless, they do represent the same line. To prove this claim, we convert to a point-slope equation: from $x = 1 + 4s$ we compute $s = x/4 - 1/4$ and substitute into y:

$$y = 4 + 2s = 4 + 2\left(\frac{x}{4} - \frac{1}{4}\right) = 4 + \frac{1}{2}x - \frac{1}{2} = \frac{1}{2}x + \frac{7}{2}$$

So, indeed, it is the same line. ▲

In our calculations in the previous example, we used $\mathbf{v}$ as the direction vector. Since $\mathbf{v}/2 = [2, 1]$ is parallel to $\mathbf{v}$, we can use it to calculate the parametric equations; we obtain (using the point A)

$$\begin{bmatrix} x \\ y \end{bmatrix} = \begin{bmatrix} -3 \\ 2 \end{bmatrix} + t_1 \begin{bmatrix} 2 \\ 1 \end{bmatrix}$$

and (using the point B)

$$\begin{bmatrix} x \\ y \end{bmatrix} = \begin{bmatrix} 1 \\ 4 \end{bmatrix} + t_2 \begin{bmatrix} 2 \\ 1 \end{bmatrix}$$

where $t_1, t_2 \in \mathbb{R}$. As before, we can convince ourselves that these sets of parametric equations yield the same point-slope equation, $y = x/2 + 7/2$.

Next, we can replace $\mathbf{v}$ by $-\mathbf{v}$, or by $14\mathbf{v}$, or by $-23\mathbf{v}$; clearly, we can build infinitely many parametric equations for a given line.

Example 4.3 Comparing Parametric Equations

(a) Find parametric equations of the line through $A = (2, 1)$ in the direction of the vector $\mathbf{v} = [-1, 4]$.

(b) Show that the parametric equations

$$\begin{cases} x = 2s \\ y = 9 - 8s \end{cases} \tag{4.8}$$

where $s \in \mathbb{R}$, also represent the line in (a).

▶ (a) The vector equation of the line is

$$\begin{bmatrix} x \\ y \end{bmatrix} = \begin{bmatrix} 2 \\ 1 \end{bmatrix} + t \begin{bmatrix} -1 \\ 4 \end{bmatrix}$$

where t is a real number. Writing out the components, we obtain the parametric equations

$$\begin{cases} x = 2 - t \\ y = 1 + 4t \end{cases} \tag{4.9}$$

(b) One way to show that the two sets of parametric equations represent the same line is to show that both yield the same point-slope equation (as we did in Example 4.2). Here is an alternative: comparing the parametric equations

$$\begin{cases} x = 2s \\ y = 9 - 8s \end{cases} \quad \text{and} \quad \begin{cases} x = 2 - t \\ y = 1 + 4t \end{cases}$$

we get

$$\begin{aligned} 2 - t &= 2s \\ 1 + 4t &= 9 - 8s \end{aligned}$$

Both equations simplify to

$$t = 2 - 2s$$

What does this mean? Substituting $s = 1$ into (4.8), we obtain the point $(x = 2, y = 1)$ on the given line. From $t = 2 - 2s$ we get that $t = 2 - 2(1) = 0$; substituting $t = 0$ into (4.9) we get $x = 2$ and $y = 1$, i.e., we get the same point. Thus, both (4.8) and (4.9) generated the same point. The formula $t = 2 - 2s$ tells us how to translate the parameter values so that the corresponding equations yield the same point.

Let's do this one more time. The value $t = 3$ substituted into (4.9) gives $x = -1$ and $y = 13$. Substituting $t = 3$ into the translation formula $t = 2 - 2s$ gives $s = -1/2$. Using (4.8) with $s = -1/2$, we obtain the same point, $x = -1$ and $y = 13$. ▲

Example 4.4 Comparing Parametric Equations

Show that the vector equations

$$\begin{bmatrix} x \\ y \end{bmatrix} = \begin{bmatrix} 3 \\ 2 \end{bmatrix} + t\begin{bmatrix} -1 \\ -1 \end{bmatrix}$$

and

$$\begin{bmatrix} x \\ y \end{bmatrix} = \begin{bmatrix} 1 \\ 0 \end{bmatrix} + s\begin{bmatrix} 2 \\ 1 \end{bmatrix}$$

where $t, s \in \mathbb{R}$, do not represent the same line. What is the relationship between the two lines?

▶ The direction vectors of the two lines, $\mathbf{v}_1 = [-1, -1]$ and $\mathbf{v}_2 = [2, 1]$, are not scalar multiples of each other. Thus, the two equations do not represent the same line, nor do they represent parallel lines. We conclude that the two lines must intersect.

To find the point of intersection, we use the strategy from Example 4.3. Combining the two sets of equations

$$\begin{cases} x = 3 - t \\ y = 2 - t \end{cases} \quad \text{and} \quad \begin{cases} x = 1 + 2s \\ y = s \end{cases}$$

we obtain

$$\begin{aligned} 3 - t &= 1 + 2s \\ 2 - t &= s \end{aligned}$$

Subtracting the second equation from the first, we get $1 = 1 + s$ and $s = 0$. It follows that $t = 2$. Using either set of parametric equations we find that $(1, 0)$ is the point of intersection of the two lines. ▲

We can compute the angle at which the two lines intersect: from

$$\begin{aligned} \cos\theta &= \frac{\mathbf{v}_1 \cdot \mathbf{v}_2}{\|\mathbf{v}_1\| \, \|\mathbf{v}_2\|} \\ &= \frac{[-1, -1] \cdot [2, 1]}{\|[-1, -1]\| \, \|[2, 1]\|} = \frac{-3}{\sqrt{2}\sqrt{5}} \approx -0.9487 \end{aligned}$$

we get that $\theta \approx 2.820$ radians (or 161.57^{o}).

Example 4.5 Converting into Parametric Equations

Find the parametric equations of the line given implicitly by $2x - 3y + 1 = 0$.

▶ Solving the given equation for x and y forces us to pick a parameter. If we let $x = t$, then

$$\begin{aligned} 2t - 3y + 1 &= 0 \\ 3y &= 2t + 1 \\ y &= \frac{2}{3}t + \frac{1}{3} \end{aligned}$$

The parametric equations are

$$\begin{cases} x = t \\ y = 2t/3 + 1/3 \end{cases}$$

where $t \in \mathbb{R}$. We conclude that the given equation represents the line through $(0, 1/3)$ whose direction is given by the vector $[1, 2/3]$ (or, whose slope is $2/3$).

Of course, we can declare y to be a parameter: $y = s$. In that case, the equation $2x - 3s + 1 = 0$ gives

$$x = \frac{3}{2}s - \frac{1}{2}$$

and the parametric equations are

$$\begin{cases} x = 3s/2 - 1/2 \\ y = s \end{cases}$$

where s is a real number. The direction vector in this case is $[3/2, 1]$; it gives a slope of $1/(3/2) = 2/3$, the same as before.

The equation $ax + by + d = 0$ represents a line in $\mathbb{R}^2$. Adding the third coordinate (variable) z, we obtain the equation $ax + by + cz + d = 0$, which *no longer represents a line.* (It is a plane, as we will soon see.)

If we wish to describe lines in $\mathbb{R}^3$, we need to use vector or parametric equations. The good news is that we use them in exactly the same way as in $\mathbb{R}^2$. Let us work through an example.

Example 4.6 Parametric Equations of a Line in Space

Find the parametric equations of the line through the points $A = (-2, 3, 0)$ and $B = (1, -2, 1)$.

▶ We use the familiar formula

$$\text{line} = \text{point} + \text{parameter} \cdot \text{direction vector}$$

Designate $A = (-2, 3, 0)$ to be the *point,* and let the direction be defined by the vector from A to B, i.e., $\mathbf{v} = [3, -5, 1]$. Thus,

$$\begin{bmatrix} x \\ y \\ z \end{bmatrix} = \begin{bmatrix} -2 \\ 3 \\ 0 \end{bmatrix} + t \begin{bmatrix} 3 \\ -5 \\ 1 \end{bmatrix}$$

or, as parametric equations:

$$\begin{cases} x = -2 + 3t \\ y = 3 - 5t \\ z = t \end{cases}$$

where t is a real number.

Everything we learned about parametric equations in $\mathbb{R}^2$ (thinking about the parameter as time, figuring out whether or not two equations represent the same line, and so on) holds in $\mathbb{R}^3$ as well.

Normal Form of the Equation of a Line in a Plane

Next, we develop another way of describing lines in $\mathbb{R}^2$. This new approach, when generalized, will help us construct the equation of a plane in $\mathbb{R}^3$.

Consider the line given implicitly by the equation

$$3x + 2y = 0$$

We can think of the left side as the dot product

$$3x + 2y = [3, 2] \cdot [x, y]$$

of the fixed vector $\mathbf{n} = [3, 2]$ and the vector $\mathbf{x} = [x, y]$ that belongs to the line; see Figure 4.4.

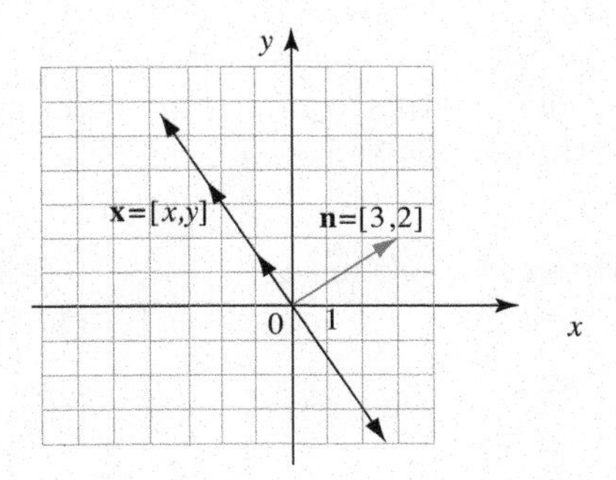

FIGURE 4.4

Normal vector to a line

Since the dot product of $\mathbf{n}$ and $\mathbf{x}$ is zero, the two vectors are orthogonal. We view the line $3x + 2y = 0$ as the collection of all vectors $\mathbf{x}$ that are perpendicular to the fixed vector $\mathbf{n}$. The vector $\mathbf{n}$ is called the *normal vector.*

In general, the equation $ax + by = 0$ can be written as

$$ax + by = [a, b] \cdot [x, y] = 0$$

or as

$$\mathbf{n} \cdot \mathbf{x} = 0$$

where $\mathbf{n} = [a, b]$ is the normal vector and $\mathbf{x} = [x, y]$ is the position vector of a point (x, y) on the line.

Using the same idea, we now construct the equation of the line in $\mathbb{R}^2$ that contains the point (x_0, y_0) and is perpendicular to a non-zero vector $\mathbf{n}$.

Let (x, y) be any point on the line, and denote by $\mathbf{v}$ the vector from (x_0, y_0) to (x, y); see Figure 4.5. Let $\mathbf{x}_0 = [x_0, y_0]$ and $\mathbf{x} = [x, y]$. Then

$$\mathbf{v} = \mathbf{x} - \mathbf{x}_0 = [x - x_0, y - y_0]$$

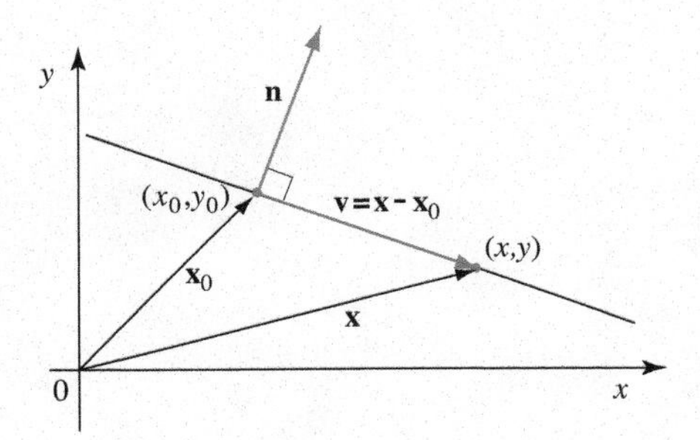

FIGURE 4.5

Normal form of the equation of a line

Since $\mathbf{n}$ is orthogonal to the line,

$$\mathbf{n} \cdot \mathbf{v} = \mathbf{n} \cdot (\mathbf{x} - \mathbf{x}_0) = 0 \tag{4.10}$$

Expanding this equation, we obtain

$$\begin{bmatrix} a \\ b \end{bmatrix} \cdot \begin{bmatrix} x - x_0 \\ y - y_0 \end{bmatrix} = 0$$

and, after calculating the dot product, we have

$$a(x - x_0) + b(y - y_0) = 0 \tag{4.11}$$

The equation (4.10), or the equivalent form (4.11), is referred to as the *normal form* of the equation of a line in $\mathbb{R}^2$.

Example 4.7 Normal Form of the Equation of a Line

The normal form of the equation of the line in a plane that contains the origin and is orthogonal to the vector $\mathbf{n} = [-3, 4]$ is $\mathbf{n} \cdot \mathbf{x} = 0$, i.e.,

$$\mathbf{n} \cdot \mathbf{x} = [-3, 4] \cdot [x, y] = 0$$

or $-3x + 4y = 0$.

The line that contains the point $\mathbf{x}_0 = [7, -2]$ and is orthogonal to $\mathbf{n} = [-3, 4]$ has the equation

$$\mathbf{n} \cdot (\mathbf{x} - \mathbf{x}_0) = 0$$

Written in coordinates,

$$\begin{bmatrix} -3 \\ 4 \end{bmatrix} \cdot \begin{bmatrix} x - 7 \\ y - (-2) \end{bmatrix} = 0$$
$$-3(x - 7) + 4(y + 2) = 0$$

i.e., $-3x + 4y + 29 = 0$.

Note that instead of calculating the dot product every time we need to use the normal equation, we can just plug the coordinates of $\mathbf{n} = [-3, 4]$ and $\mathbf{x}_0 = [7, -2]$ into $a(x - x_0) + b(y - y_0) = 0$.

Equation of a Plane in Space

Fix a non-zero vector $\mathbf{n} = [a, b, c]$ in $\mathbb{R}^3$. The vectors orthogonal to $\mathbf{n}$ no longer form a line, as they did in two dimensions. Instead, there are many more of them, pointing in different directions; see Figure 4.6. As the figure suggests, these vectors generate a collection of parallel planes. To single out one of those planes, we impose an additional requirement, namely that the plane contains a chosen point $X_0 = (x_0, y_0, z_0)$. (In other words, there is a unique plane containing the given point that is perpendicular to the given vector.)

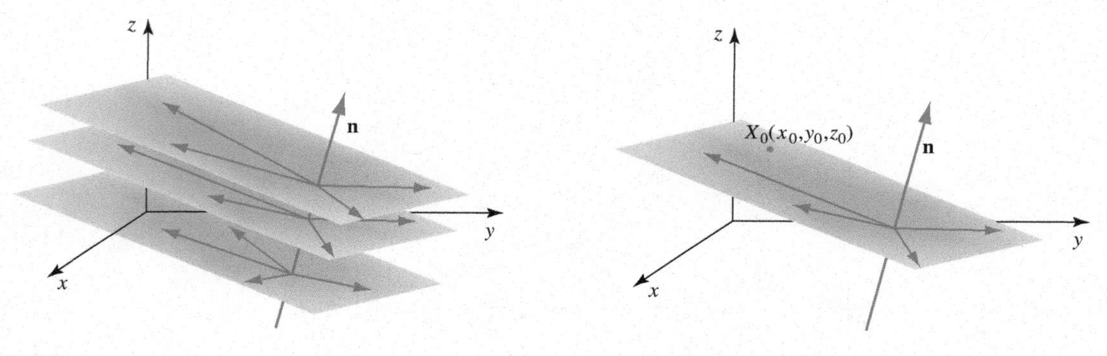

FIGURE 4.6
Vectors orthogonal to a fixed vector generate a collection of parallel planes

Let $X = (x, y, z)$ be a point in the plane determined by $X_0 = (x_0, y_0, z_0)$ and $\mathbf{n}$. The vector

$$\mathbf{v} = \mathbf{x} - \mathbf{x}_0 = [x - x_0, y - y_0, z - z_0]$$

from X to X_0 lies in the plane, and is therefore orthogonal to $\mathbf{n}$; see Figure 4.7. (Since we can move vectors around, we placed $\mathbf{n}$ so that its tail is located at X_0.) It follows that

$$\mathbf{n} \cdot \mathbf{v} = \mathbf{n} \cdot (\mathbf{x} - \mathbf{x}_0) = 0 \qquad (4.12)$$

(note that this equation is of exactly the same form as the normal equation of a line in $\mathbb{R}^2$). Equation (4.12) is called the *normal form* of the equation of a plane. Written in coordinates, the equation reads

$$\begin{bmatrix} a \\ b \\ c \end{bmatrix} \cdot \begin{bmatrix} x - x_0 \\ y - y_0 \\ z - z_0 \end{bmatrix} = 0$$

and, after we calculate the dot product,

$$a(x - x_0) + b(y - y_0) + c(z - z_0) = 0$$

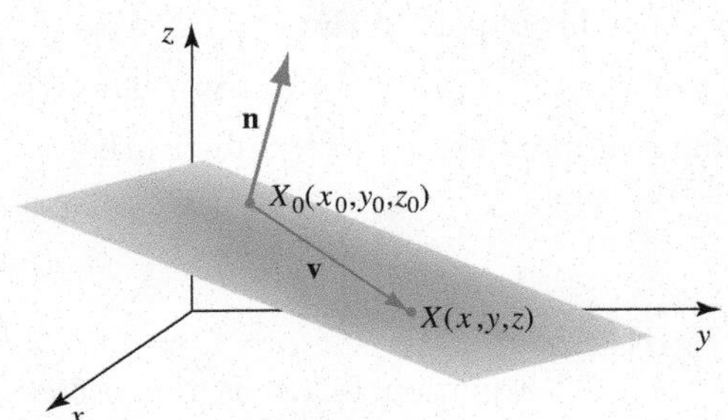

FIGURE 4.7

The plane containing a given point and orthogonal to a given vector

Example 4.8 Equation of a Plane in Space

Find the equation of the plane in $\mathbb{R}^3$ through the point $(3, 0, -2)$ that is perpendicular to the vector $[-1, 2, 11]$.

▶ Using the equation we just discovered,

$$(-1)(x - 3) + 2(y - 0) + 11(z - (-2)) = 0$$

Simplifying, we get $-x + 2y + 11z + 25 = 0$. ▲

We can characterize the xy-plane as the plane through the origin that is orthogonal to the unit vector $\mathbf{k} = [0, 0, 1]$. Thus, its equation is

$$0(x - 0) + 0(y - 0) + 1(z - 0) = 0$$

i.e., $z = 0$. The plane parallel to the xy-plane (i.e., orthogonal to $\mathbf{k} = [0, 0, 1]$) through the point $(3, 4, -7)$ has the equation

$$0(x - 3) + 0(y - 4) + 1(z - (-7)) = 0$$

i.e., $z = -7$.

Because we will need them later, we construct the *parametric equations* of a plane.

In order to do so, we observe that a plane is uniquely determined by specifying one point $X_0 = (x_0, y_0, z_0)$ and two non-parallel, non-zero vectors $\mathbf{u}$ and $\mathbf{v}$ that lie in it. (Instead of the colloquial "vectors $\mathbf{u}$ and $\mathbf{v}$ lie in a plane," we could say that $\mathbf{u}$ and $\mathbf{v}$ are parallel to the plane, or that $\mathbf{u}$ and $\mathbf{v}$ are orthogonal to the normal vector associated with the plane.)

For convenience, we place the tails of $\mathbf{u}$ and $\mathbf{v}$ at X_0; see Figure 4.8.

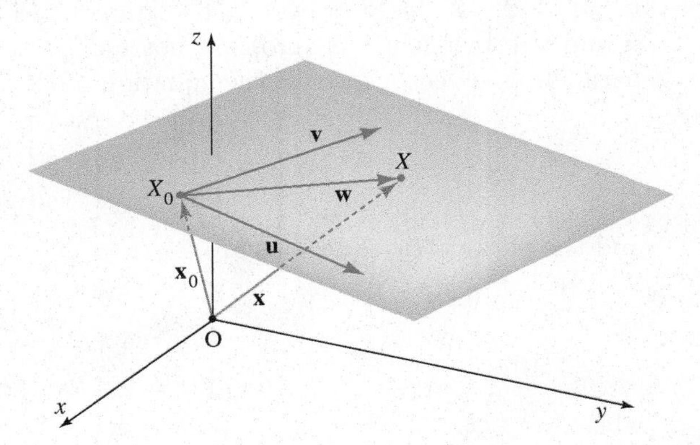

FIGURE 4.8

Deriving parametric equations of a plane

Let $X = (x, y, z)$ be a point in the plane, and denote by $\mathbf{x}$ and $\mathbf{x}_0$ the position vectors of X and X_0. The vector $\mathbf{w} = \mathbf{x} - \mathbf{x}_0$ (the vector from X_0 to X) lies in the same plane as $\mathbf{u}$ and $\mathbf{v}$.

The parallelogram law (Figure 4.9) tells us that $\mathbf{w}$ can be expressed as the sum of a vector parallel to $\mathbf{u}$ (i.e., a scalar multiple of $\mathbf{u}$) and a vector parallel to $\mathbf{v}$ (a scalar multiple of $\mathbf{v}$); thus,

$$\mathbf{w} = s\mathbf{u} + t\mathbf{v}$$

for some real numbers s and t. Substituting $\mathbf{w} = \mathbf{x} - \mathbf{x}_0$, we get

$$\mathbf{x} - \mathbf{x}_0 = s\mathbf{u} + t\mathbf{v}$$

and

$$\mathbf{x} = \mathbf{x}_0 + s\mathbf{u} + t\mathbf{v} \tag{4.13}$$

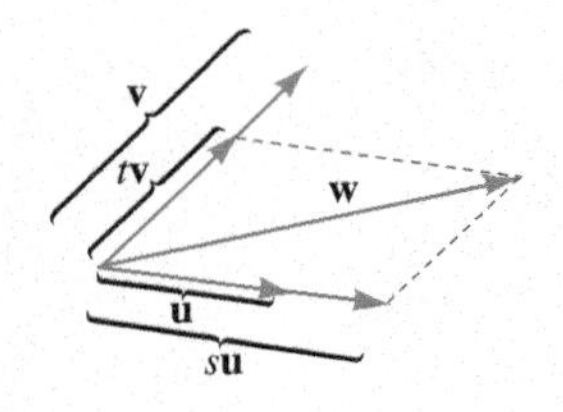

FIGURE 4.9

Expressing $\mathbf{w}$ in terms of $\mathbf{u}$ and $\mathbf{v}$

where $s, t \in \mathbb{R}$. Equation (4.13) is the *vector form* of the equation of a plane. Recall that $\mathbf{x}_0 = [x_0, y_0, z_0]$ is the position vector of the point in the plane, and $\mathbf{u} = [u_1, u_2, u_3]$ and $\mathbf{v} = [v_1, v_2, v_3]$ are the vectors in the plane (assumed non-zero, and not parallel to each other).

In components,

$$\begin{bmatrix} x \\ y \\ z \end{bmatrix} = \begin{bmatrix} x_0 \\ y_0 \\ z_0 \end{bmatrix} + s \begin{bmatrix} u_1 \\ u_2 \\ u_3 \end{bmatrix} + t \begin{bmatrix} v_1 \\ v_2 \\ v_3 \end{bmatrix}$$

The corresponding set of equations

$$\begin{cases} x = x_0 + su_1 + tv_1 \\ y = y_0 + su_2 + tv_2 \\ z = z_0 + su_3 + tv_3 \end{cases} \tag{4.14}$$

where $s, t \in \mathbb{R}$, are the *parametric equations* of a plane in space.

Note that we need two parameters to describe a plane; we needed only one for a line. This is one reason why we say that a line is one-dimensional, whereas a plane is said to have two dimensions.

Example 4.9 Parametric Equations of a Plane

Find the parametric equations of the plane that contains the points $A = (0, 4, 2)$, $B = (-1, 6, 0)$, and $C = (3, -3, 4)$.

▶ We are looking for an equation of the form

$$\begin{aligned} \text{plane} &= \text{point} + \text{parameter} \cdot \text{vector} \\ &\quad + \text{second parameter} \cdot \text{second vector} \end{aligned} \tag{4.15}$$

where the two vectors must be non-zero and not parallel.

Designate $A = (0, 4, 2)$ to be the *point.* Let $\mathbf{u}$ be the vector from A to B, i.e., $\mathbf{u} = [-1, 2, -2]$; let $\mathbf{v}$ be the vector from A to C, i.e., $\mathbf{v} = [3, -7, 2]$. The vectors $\mathbf{u}$ and $\mathbf{v}$ (obviously non-zero) are not multiples of each other and hence are not parallel. The vector form of the equation of the plane is

$$\begin{bmatrix} x \\ y \\ z \end{bmatrix} = \begin{bmatrix} 0 \\ 4 \\ 2 \end{bmatrix} + s \begin{bmatrix} -1 \\ 2 \\ -2 \end{bmatrix} + t \begin{bmatrix} 3 \\ -7 \\ 2 \end{bmatrix}$$

and the parametric equations are

$$\begin{cases} x = -s + 3t \\ y = 4 + 2s - 7t \\ z = 2 - 2s + 2t \end{cases}$$

where $s, t \in \mathbb{R}$. ▲

Note that if we need to convert the parametric equations of Example 4.9 to the implicit form $ax + by + cz + d = 0$, we have to eliminate both parameters (see Exercise 33).

Example 4.10 Parametric Equation of a Plane

In this example, we use the fact that a plane is uniquely determined by two intersecting lines.

Find the vector equation of the plane that contains the lines

$$[x, y, z] = [1, 1, 4] + t_1[0, -2, 1]$$

and

$$[x, y, z] = [1, 1, 4] + t_2[2, 0, 1]$$

with $t_1, t_2 \in \mathbb{R}$.

▶ Clearly, the two lines intersect at the point whose position vector is $\mathbf{x}_0 = [1, 1, 4]$. Thus, we have one ingredient for (4.15), namely the *point;* we need two non-parallel vectors. We actually have those as well, no calculations needed—the direction vectors of the two lines belong to the plane. So let $\mathbf{u} = [0, -2, 1]$ and $\mathbf{v} = [2, 0, 1]$. The required equation is

$$\mathbf{x} = \mathbf{x}_0 + s\mathbf{u} + t\mathbf{v}$$

or

$$\begin{bmatrix} x \\ y \\ z \end{bmatrix} = \begin{bmatrix} 1 \\ 1 \\ 4 \end{bmatrix} + s \begin{bmatrix} 0 \\ -2 \\ 1 \end{bmatrix} + t \begin{bmatrix} 2 \\ 0 \\ 1 \end{bmatrix}$$

where s and t are real numbers. ▲

Summary We studied a number of ways of describing lines and planes. (In the list that follows, we assume that the parameters s and t are real numbers.)

(a) Line in $\mathbb{R}^2$:

Implicit form: $ax + by + d = 0$

Vector form (of the parametric equation): $\mathbf{p} = \mathbf{a} + t\mathbf{v}$; $\mathbf{a}$ is a point on the line and $\mathbf{v}$ is the direction vector

Parametric equation(s):

$$\begin{cases} x = a_1 + tv_1 \\ y = a_2 + tv_2 \end{cases}$$

$\mathbf{a} = [a_1, a_2]$ is a point on the line and $\mathbf{v} = [v_1, v_2]$ is the direction vector

Normal form: $\mathbf{n} \cdot (\mathbf{x} - \mathbf{x}_0) = 0$ or $a(x - x_0) + b(y - y_0) = 0$; $\mathbf{x}_0 = (x_0, y_0)$ is a point on the line, $\mathbf{x} = [x, y]$, and and $\mathbf{n} = [a, b]$ is the normal vector (orthogonal to the line)

(b) Line in $\mathbb{R}^3$:

Vector form (of the parametric equation): $\mathbf{p} = \mathbf{a} + t\mathbf{v}$; $\mathbf{a}$ is a point on the line and $\mathbf{v}$ is the direction vector

Parametric equation(s):

$$\begin{cases} x = a_1 + tv_1 \\ y = a_2 + tv_2 \\ z = a_3 + tv_3 \end{cases}$$

$\mathbf{a} = [a_1, a_2, a_3]$ is a point on the line and $\mathbf{v} = [v_1, v_2, v_3]$ is the direction vector

(c) Plane in $\mathbb{R}^3$:

Implicit form: $ax + by + cz + d = 0$

Vector form (of the parametric equation): $\mathbf{p} = \mathbf{x}_0 + s\mathbf{u} + t\mathbf{v}$; $\mathbf{x}_0$ is a point in the plane and $\mathbf{u}$ and $\mathbf{v}$ lie in the plane

Parametric equation(s):

$$\begin{cases} x = x_0 + su_1 + tv_1 \\ y = y_0 + su_2 + tv_2 \\ z = z_0 + su_3 + tv_3 \end{cases} \tag{4.16}$$

$\mathbf{x}_0 = [x_0, y_0, z_0]$ is a point in the plane and the vectors $\mathbf{u} = [u_1, u_2, u_3]$ and $\mathbf{v} = [v_1, v_2, v_3]$ lie in the plane

Normal form: $\mathbf{n} \cdot (\mathbf{x} - \mathbf{x}_0) = 0$ or $a(x - x_0) + b(y - y_0) + c(z - z_0) = 0$; $\mathbf{x}_0 = (x_0, y_0, z_0)$ is a point in the plane, $\mathbf{x} = [x, y, z]$, and $\mathbf{n} = [a, b, c]$ is the normal vector (orthogonal to the plane)

4 Exercises

1. Find a direction vector of any line of slope 4.
2. Find a direction vector of the line $y = -2x - 1$.
3. What is the slope of a line whose direction vector is $\mathbf{v} = [v_1, v_2]$? (Hint: What *is* the slope of a line?)
4. Write the parametric equations of the y-axis in $\mathbb{R}^2$.
5. Write the parametric equations of the y-axis in $\mathbb{R}^3$.
6. Write the parametric equations of the yz-plane in $\mathbb{R}^3$.
7. Write the parametric equations of the xz-plane in $\mathbb{R}^3$.
8. Find the vector and parametric equations of the line given implicitly by $3x - 7y + 1 = 0$.
9. Find the vector and parametric equations of the line given implicitly by $x - 5y + 9 = 0$.

10–17 ▪ In each case, find the required equation(s) of a line.

10. Find the parametric equations of the line passing through the points $(-1, -2)$ and $(0, 5)$ in $\mathbb{R}^2$. What is the slope of the line?
11. Find the parametric equations of the line passing through the points $(4, 0)$ and $(0, -2)$ in $\mathbb{R}^2$. Convert the equations into implicit form.
12. Find the parametric equations of the line passing through the origin and the point $(1, -2, -3)$ in $\mathbb{R}^3$.
13. Find the parametric equations of the line passing through the points $(0, 5, 0)$ and $(-1, 8, 2)$ in $\mathbb{R}^3$.
14. Find the vector and parametric equations of the line passing through $(3, -4)$ with direction vector $[5, 8]$. What is the slope of the line?
15. Find the vector and parametric equations of the line passing through $(1, 2, 7)$ with direction vector $[-2, 0, 4]$.
16. Find the normal and implicit forms of the equation of the line through $(2, 9)$ with normal vector $[-3, -2]$.
17. Find the normal and parametric equations of the line through $(-4, 6)$ with normal vector $[1, 1]$.

18. Pick any two points on the line $\mathbf{p} = \mathbf{a} + t\mathbf{v}$ (this means that we need to pick two different times, t_1 and t_2, and consider corresponding points $\mathbf{p}_1 = \mathbf{a} + t_1\mathbf{v}$ and $\mathbf{p}_2 = \mathbf{a} + t_2\mathbf{v}$). Compute the speed as distance/time. Show that the speed is equal to $\|\mathbf{v}\|$.

19. Find the parametric equations of the line that passes through $(3, -1)$ and is perpendicular to the line $2x - 5y - 4 = 0$.

20. Find the parametric equations of the line that passes through $(3, -1)$ and is parallel to the line $2x - 5y - 4 = 0$.

21–30 ▪ In each case, find the required equation(s) of a plane.

21. Find the vector and parametric equations of the plane containing the point $(3, 4, -1)$ and parallel to the vectors (i.e., containing the vectors) $\mathbf{v} = [0, 1, 4]$ and $\mathbf{w} = [0, 0, 2]$.

22. Find the vector and parametric equations of the plane containing the point $(2, 7, 0)$ and parallel to the vectors $\mathbf{v} = 2\mathbf{i} - \mathbf{j} + \mathbf{k}$ and $\mathbf{w} = 4\mathbf{j} + \mathbf{k}$.

23. Find the vector and parametric equations of the plane passing through the points $(1, 3, 4)$, $(3, 9, -6)$, and $(1, 0, 0)$.

24. Find the parametric and implicit equations of the plane containing the points $(4, 0, 0)$, $(0, 2, 0)$, and $(0, 0, -3)$.

25. Find the parametric and implicit equations of the plane containing the point $(4, 2, -5)$ with normal vector $\mathbf{n} = 3\mathbf{i} + \mathbf{k}$.

26. Find the vector and implicit equations of the plane containing the point $(0, 4, 0)$ with normal vector $\mathbf{n} = -5\mathbf{i} + 2\mathbf{j} + 2\mathbf{k}$.

27. Find the parametric equations of the plane passing through $(2, 1, 3)$ parallel to the plane $x + 4z - 1 = 0$.

28. Find the parametric equations of the plane passing through $(0, 0, 1)$ that is parallel to the plane $x + 2y + 7z - 1 = 0$.

29. Find the parametric equations of the plane that contains the two intersecting lines

$$\begin{bmatrix} x \\ y \\ z \end{bmatrix} = \begin{bmatrix} 0 \\ 2 \\ -6 \end{bmatrix} + t_1 \begin{bmatrix} 0 \\ 4 \\ 2 \end{bmatrix} \quad \text{and} \quad \begin{bmatrix} x \\ y \\ z \end{bmatrix} = \begin{bmatrix} 0 \\ 2 \\ -6 \end{bmatrix} + t_2 \begin{bmatrix} 1 \\ -1 \\ 2 \end{bmatrix}$$

where $t_1, t_2 \in \mathbb{R}$.

30. Find the parametric equations of the plane that contains the two intersecting lines

$$\begin{bmatrix} x \\ y \\ z \end{bmatrix} = \begin{bmatrix} 0 \\ -1 \\ 2 \end{bmatrix} + t_1 \begin{bmatrix} 1 \\ 1 \\ 2 \end{bmatrix} \quad \text{and} \quad \begin{bmatrix} x \\ y \\ z \end{bmatrix} = \begin{bmatrix} 2 \\ 1 \\ 6 \end{bmatrix} + t_2 \begin{bmatrix} 0 \\ -1 \\ 4 \end{bmatrix}$$

where $t_1, t_2 \in \mathbb{R}$.

31. Convert the implicit equation of the plane $3x - y - 2z + 4 = 0$ into parametric equations.

32. Convert the implicit equation of the plane $-5x - y + 2z + 1 = 0$ into vector and parametric equations.

33. Convert the parametric equations of Example 4.9 into implicit form. (Hint: Take any two equations (say, the first two) and solve for s and t. In other words, express s and t in terms of x and y. Substitute these expressions into the remaining equation and simplify.)

34. Find the vector equation of the line that passes through $(1, 2, -2)$ and is perpendicular to the plane given by $4y - z - 3 = 0$.

35. Determine whether or not the planes $x - 2y + 8z - 4 = 0$ and $2x - 4y + 8z - 4 = 0$ are parallel.

36. Show that the planes $4x - y + z - 3 = 0$ and $3y + 3z + 7 = 0$ are perpendicular.

37. Find the parametric equations of the line that is the intersection of the planes $x + 3z - 1 = 0$ and $2x + y - 3z - 6 = 0$.

38. Find the parametric equations of the line that is the intersection of the planes $x + y - 1 = 0$ and $2x - z - 2 = 0$.

39. Find the parametric equations of the line that passes through $(0, 8, 2)$ and is perpendicular to the plane given implicitly by $x - 3y + z - 4 = 0$.

40. Determine whether the line

$$\begin{bmatrix} x \\ y \\ z \end{bmatrix} = \begin{bmatrix} 1 \\ 2 \\ -6 \end{bmatrix} + t \begin{bmatrix} 1 \\ 1 \\ -3 \end{bmatrix}$$

$(t \in \mathbb{R})$ and the plane $2x + y + z + 4 = 0$ are parallel or perpendicular.

41. Determine whether the line

$$\begin{bmatrix} x \\ y \\ z \end{bmatrix} = \begin{bmatrix} 1 \\ 0 \\ 0 \end{bmatrix} + t \begin{bmatrix} 4 \\ 1 \\ -1 \end{bmatrix}$$

$(t \in \mathbb{R})$ and the plane $-8x - 2y + 2z + 3 = 0$ are parallel or perpendicular.

42. Show that the parametric equations

$$\begin{cases} x = \alpha t \\ y = 3\alpha t/2 \end{cases}$$

where $t \in \mathbb{R}$, represent the same line, no matter what value of $\alpha \neq 0$ we take. Show that the parametric equations

$$\begin{cases} x = 2\beta s \\ y = 3\beta s \end{cases}$$

where $s \in \mathbb{R}$, represent the same line as the first set of equations (as long as $\beta \neq 0$).

5 Systems of Linear Equations

Why do we need systems of linear equations?

Consider an example. Two animal populations, A and B, share a region within an ecosystem. In the absence of the other species, each population increases in size. Assuming that the per capita growth rates are a and b, respectively, we write

$$\frac{dn_A}{dt} = an_A$$
$$\frac{dn_B}{dt} = bn_B$$

where n_A and n_B denote the number of individuals in populations A and B, and $a, b > 0$. When both species are present in the region, the competition for space and resources causes a decline in their growth rates. Assuming that the decline in the growth rate of one population is proportional to the size of the other, we obtain the system

$$\begin{aligned} \frac{dn_A}{dt} &= an_A - cn_B \\ \frac{dn_B}{dt} &= bn_B - dn_A \end{aligned} \tag{5.1}$$

where $c, d > 0$. The differential equations in (5.1) represent the simplest model of a competitive relationship between two species.

For various reasons, it is important to find the *equilibrium* of a system. In this case, the equilibrium consists of the values of n_A and n_B that remain unchanged under the dynamics given by the system (5.1). This means that when a member of a population dies or is killed, it is immediately replaced by another member, so that the *number* of individuals does not change.

To find the equilibrium, we need to solve the system $dn_A/dt = 0$, $dn_B/dt = 0$:

$$\begin{aligned} an_A - cn_B &= 0 \\ bn_B - dn_A &= 0 \end{aligned} \tag{5.2}$$

The equations in (5.2) form a system of two linear equations with two variables, and its solutions give the desired equilibrium.

Thus, we solve linear equations when we need to find the equilibrium of a system of linear differential equations. As we will see in Section 7, in order for a computed tomography (CT) scan to produce an image on a screen, a large linear system needs to be solved. Linear equations help us find eigenvectors, which are an important algebraic tool. For instance, we use eigenvectors to solve a system such as (5.1) in Example 11.6 at the end of Section 11. We use eigenvectors again in Section 12.

Linear Systems

We met linear equations in Section 4: the equation $ax + by + d = 0$ represents a line in $\mathbb{R}^2$, and $ax + by + cz + d = 0$ describes a plane in $\mathbb{R}^3$.

Definition 14 **Linear Equation**

A *linear equation* in variables $x_1, x_2, \ldots, x_n$ is an equation of the form

$$a_1x_1 + a_2x_2 + \cdots + a_nx_n = b$$

where $a_1, a_2, \ldots, a_n$ and b are real numbers.

Very often, when the number of variables is small, we do not use subscripts. Instead, we pick letters from the end of the alphabet: x and y (in the case of two variables); x, y, and z (three variables); or x, y, z, and w (four variables). The real numbers $a_1, a_2, \ldots, a_n$ are called the *coefficients* and b is called the *constant term* or the *free coefficient*.

Example 5.1 Linear and Non-linear Equations

The equation

$$3x_1 - 0.2x_2 - 3x_4 + 1.5x_4 = 10$$

is a linear equation in variables x_1, x_2, x_3, and x_4. The equation

$$x - y - 5z = -4$$

is a linear equation in variables x, y, and z, and

$$x/2 - \sqrt{5}y = 11$$

is a linear equation in x and y.

A linear equation can only contain linear functions of its variables. If an equation contains any other function of any of its variables, then it is not linear (also called *non-linear*). The following equations are not linear:

$$3x^2 - y + 7z = -4$$
$$e^{x_1} + x_2 = 1$$
$$xy + 4z = 10$$

Definition 15 Solution of an Equation

A solution of a linear equation

$$a_1x_1 + a_2x_2 + \cdots + a_nx_n = b \tag{5.3}$$

is any set of values $x_1 = s_1$, $x_2 = s_2$, $\ldots$, $x_n = s_n$ that satisfy the given equation; i.e., when substituted into (5.3), the values $x_1 = s_1$, $x_2 = s_2$, $\ldots$, $x_n = s_n$ give an identity.

For example, $x = 3$ and $y = 4$ form a solution of the equation $3x - y = 5$, since

$$3x - y = 5$$
$$3(3) - (4) = 5$$
$$5 = 5$$

Likewise, if $x = 5/3$ and $y = 0$, then $3x - y = 3(5/3) - (0) = 5$. On the other hand, the values $x = -2$ and $y = 4$ do not belong to the solution set of $3x - y = 5$, since $3x - y = 3(-2) - (4) = -10 \neq 5$.

Depending on the context, we write a solution as a list

$$x_1 = s_1, x_2 = s_2, \ldots, x_n = s_n$$

or as an ordered n-tuple

$$(s_1, s_2, \ldots, s_n)$$

or as a vector

$$[s_1, s_2, \ldots, s_n]$$

Thus, $(3, 4)$ and $(5/3, 0)$, or the vectors $[3, 4]$ and $[5/3, 0]$, are solutions of the equation $3x - y = 5$.

To *solve an equation* means to find *all* solutions.

From Section 4 we know that the equation $3x - y = 5$ represents a line, which contains infinitely many points. Thus, the solution set of this equation is infinite.

If we let $x = t$, then $y = 3x - 5 = 3t - 5$, and the solution is

$$\{(t, 3t - 5) \mid t \in \mathbb{R}\}$$

or, using vector notation,

$$\begin{bmatrix} x \\ y \end{bmatrix} = \begin{bmatrix} t \\ 3t - 5 \end{bmatrix} = \begin{bmatrix} 0 \\ -5 \end{bmatrix} + t\begin{bmatrix} 1 \\ 3 \end{bmatrix}$$

($t \in \mathbb{R}$), which we recognize as the vector equation of the line $3x - y = 5$.

Definition 16 System of Linear Equations

An $m \times n$ *system of linear equations* (or an $m \times n$ *linear system*) is a set of m linear equations, each containing the same variables, $x_1, x_2, \ldots, x_n$:

$$\begin{aligned} a_{11}x_1 + a_{12}x_2 + \cdots + a_{1n}x_n &= b_1 \\ a_{21}x_1 + a_{22}x_2 + \cdots + a_{2n}x_n &= b_2 \\ &\cdots \\ a_{m1}x_1 + a_{m2}x_2 + \cdots + a_{mn}x_n &= b_m \end{aligned}$$

A *solution* of a system is a set of values $x_1 = s_1, x_2 = s_2, \ldots, x_n = s_n$, that satisfy *all* equations. The collection of all solutions of a given system of equations is called the *solution set*.

The coefficients a_{ij} have two subscripts: the first identifies the equation (1 for the first, 2 for the second, ..., m for the mth equation), and the second identifies the variable (all coefficients with a 1 in the second subscript are the coefficients of x_1, all coefficients with a 2 in the second subscript are the coefficients of x_2, and so on). It is assumed that at least one of the coefficients a_{ij} in each row is not zero. If all constant terms (free coefficients) $b_1, b_2, \ldots, b_m$ are zero, the system is called *homogeneous*.

As in the case of a single equation, we can write a solution of the system of equations in the form of a list (as in Definition 16), as an ordered n-tuple, or as a vector in $\mathbb{R}^n$.

To warm up, we study 2×2 linear systems first.

Systems of Two Equations in Two Variables

A 2×2 linear system in variables x_1 and x_2 is given by

$$\begin{aligned} a_{11}x_1 + a_{12}x_2 &= b_1 \\ a_{21}x_1 + a_{22}x_2 &= b_2 \end{aligned} \tag{5.4}$$

Replacing the variables x_1 and x_2 by x and y, and dropping subscripts, we write the system (5.4) as

$$\begin{aligned} Ax + By &= P \\ Cx + Dy &= Q \end{aligned} \tag{5.5}$$

where A, B, C, D, P, and Q are real numbers.

Because the system is small (two equations, two variables), we are able to drop subscripts and abandon the general form (5.4). However, as the system starts growing, we run into problems: for a 3×3 system, we need 9 symbols for the coefficients, 3 for the variables, and 3 for the constant terms, for a total of 15 letters. A 4×4 system exhausts almost the whole alphabet, requiring 24 different letters. So for larger systems we use the subscript notation introduced in Definition 16.

The solution set of a single linear equation in two variables is a line. Thus, each equation in (5.5) represents one line. Given that there are three possibilities for the mutual position of two lines in a plane, we obtain the following result.

Theorem 4 Solution Set of a 2×2 Linear System

The solution set of a 2×2 linear system is determined by the mutual position of the two lines it represents. One of the following three cases holds:

(a) The two lines intersect at exactly one point. The system (5.5) has a unique solution (Figure 5.1a).

(b) The two lines are parallel and intersect, i.e., have all points in common. The system (5.5) has infinitely many solutions (Figure 5.1b).

(c) The two lines are parallel and do not intersect. The system (5.5) has no solutions (Figure 5.1c).

If (a) or (b) occurs, we say that the system (5.5) is *consistent.* (In general, a system is called *consistent* if it has at least one solution.) A system that has no solutions is called *inconsistent.* A system represented by two parallel, non-intersecting lines (case (c) in Theorem 4) is an example of an inconsistent system.

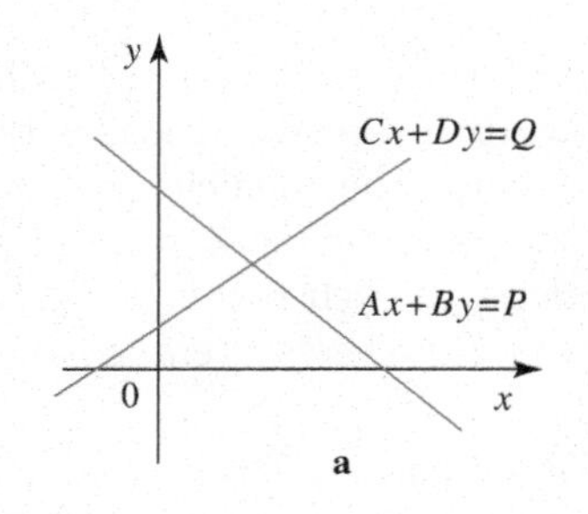

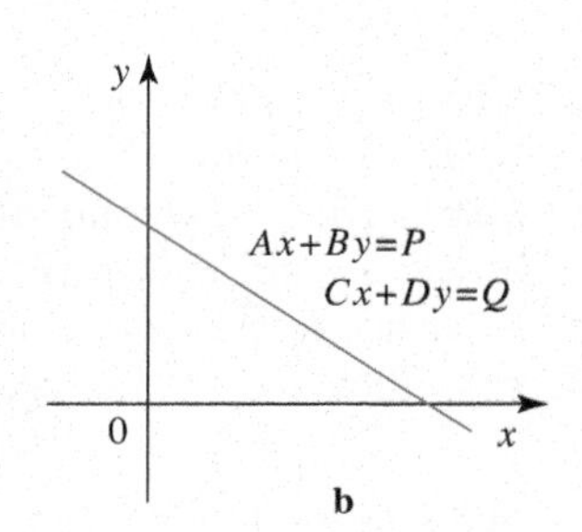

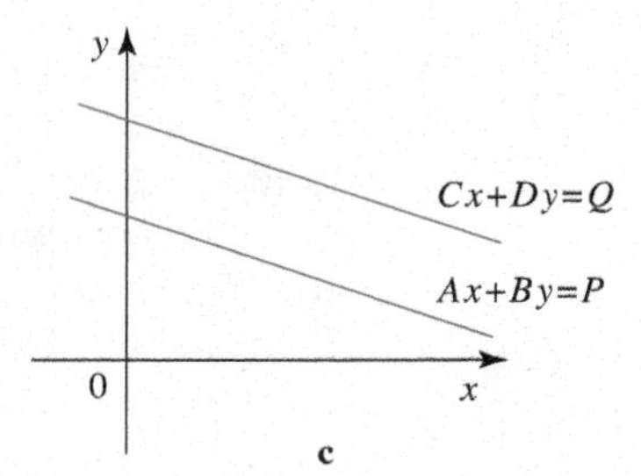

FIGURE 5.1
Mutual positions of two lines in a plane

Thus, a 2×2 system has a unique solution, or infinitely many solutions, or no solutions at all. We now explore examples illustrating all three cases.

Example 5.2 A System with a Unique Solution

Solve the system

$$\begin{aligned} -4x + 3y &= 13 \\ 2x + y &= 1 \end{aligned}$$

Solving each equation for y, we get

$$\begin{aligned} y &= \frac{4}{3}x + \frac{13}{3} \\ y &= -2x + 1 \end{aligned} \tag{5.6}$$

Clearly, the two lines have different slopes; since they are not parallel, they must intersect at a single point. Combining equations (5.6), we obtain

$$\begin{aligned} \frac{4}{3}x + \frac{13}{3} &= -2x + 1 \\ \frac{10}{3}x &= -\frac{10}{3} \\ x &= -1 \end{aligned}$$

Using either equation in (5.6), we obtain $y = 3$. In conclusion, $x = -1$, $y = 3$ is the only solution of the given system.

An alternative way to solve the system

$$-4x + 3y = 13$$
$$2x + y \ = \ 1$$

is to multiply the second equation by 2 and add to the first, thus getting

$$5y = 15$$

and $y = 3$. From either equation, we get $x = -1$.

Example 5.3 **A System with Infinitely Many Solutions**

Solve the system

$$3x - 2y = 4$$
$$-6x + 4y = -8$$

▶ Solving the first equation for y, we get $2y = 3x - 4$ and

$$y = \frac{3}{2}x - 2$$

From the second equation, we get $4y = 6x - 8$ and (after dividing by 4)

$$y = \frac{3}{2}x - 2$$

Thus, both equations represent the same line. (Another way to reach the same conclusion is to realize that by multiplying the first equation by -2 we obtain the second equation.) We conclude that the given system has infinitely many solutions, represented by the points on the line $3x - 2y = 4$.

A convenient way to describe the solutions in this case is to use a parameter. Let $x = t$; then $3t - 2y = 4$, and $y = 3t/2 - 2$. We write the solution set as

$$\{(t, 3t/2 - 2) \mid t \in \mathbb{R}\}$$

▲

Example 5.4 **A System with No Solutions**

Solve the system

$$3x - y = 7$$
$$6x - 2y = 8$$

▶ Solving both equations for y, we obtain $y = 3x - 7$ and $y = 3x - 4$. The two lines are parallel, and since their y-intercepts are different (-7 and -4), we conclude that they have no points in common. Consequently, the given system has no solutions.

To see this algebraically, we combine the two equations, obtaining

$$3x - 7 = 3x - 4$$
$$-7 = -4$$

which is obviously wrong. Thus, there are no points that satisfy both equations, and the system has no solutions.

▲

Let's go back to the system

$$-4x + 3y = 13$$
$$2x + y = 1$$

from Example 5.2. Using vectors and vector operations, we rewrite the two equations in the following way:

$$\begin{bmatrix} -4x + 3y \\ 2x + y \end{bmatrix} = \begin{bmatrix} 13 \\ 1 \end{bmatrix}$$

$$\begin{bmatrix} -4x \\ 2x \end{bmatrix} + \begin{bmatrix} 3y \\ y \end{bmatrix} = \begin{bmatrix} 13 \\ 1 \end{bmatrix}$$

$$x\begin{bmatrix} -4 \\ 2 \end{bmatrix} + y\begin{bmatrix} 3 \\ 1 \end{bmatrix} = \begin{bmatrix} 13 \\ 1 \end{bmatrix} \tag{5.7}$$

The components of the two vectors on the left side are the coefficients of x and y, and the components of the vector on the right side are the constant terms (free coefficients). The expression on the left side in (5.7) is a linear combination of the vectors

$$\begin{bmatrix} -4 \\ 2 \end{bmatrix} \quad \text{and} \quad \begin{bmatrix} 3 \\ 1 \end{bmatrix}$$

From this point of view, solving the given system of equations amounts to expressing the vector of constant terms

$$\begin{bmatrix} 13 \\ 1 \end{bmatrix}$$

as a linear combination of the vectors carrying the coefficients of the two variables.

The comment following Example 5.2 suggests that there are various ways of solving a 2×2 linear system. As well, our ability to visualize the equations as lines and then argue geometrically helps a great deal.

Next, we work on a method that can be used to solve larger linear systems, for example a system of three equations in four variables, or a system of two equations in five variables. Unfortunately, we can no longer rely on geometry to visualize the equations (solving a 3×4 system requires that we work in four dimensions).

A General Method for Solving Linear Systems

Consider the $m \times n$ system (i.e., the system of m equations in n variables)

$$\begin{aligned} a_{11}x_1 + a_{12}x_2 + \cdots + a_{1n}x_n &= b_1 \\ a_{21}x_1 + a_{22}x_2 + \cdots + a_{2n}x_n &= b_2 \\ &\cdots \\ a_{m1}x_1 + a_{m2}x_2 + \cdots + a_{mn}x_n &= b_m \end{aligned}$$

where $m, n \geq 2$. Before we start looking for solutions, it would be good to know what we are looking for. As in the 2×2 case, there are three possibilities.

Theorem 5 Solutions of a Linear System

Any system of linear equations has a unique solution, or infinitely many solutions, or no solutions at all.

A system that has at least one solution is called *consistent*. An *inconsistent* system has no solutions. Theorem 5 says that a consistent linear system exhibits extremes: it has either the smallest number of solutions (one) or the largest number (infinitely many). A linear system cannot have exactly two, three, or ten solutions.

How do we solve a general linear system?

The idea is fairly straightforward and relies on the work we have done already: we convert the given linear system into a new linear system that

(a) has the same solutions as the original system and

(b) is easier to solve.

Definition 17 Equivalent Linear Systems

Two linear systems are called *equivalent* if they have the same solution set.

In order to obtain an equivalent system from a given system, we can

(a) multiply an equation by a non-zero real number

(b) add two equations

(c) rearrange the order of the equations

Now we know what we are allowed to do. But what linear systems are easy, or easier, to solve? Consider an example.

Example 5.5 Solving a System by Reduction to an Equivalent System

Solve the system

$$\begin{aligned} -2x + 3y &= 5 \qquad (R_1) \\ 3x + 9y &= -3 \qquad (R_2) \end{aligned}$$

In order to keep track of calculations, we label the equations by R_1, R_2, and so on (R stands for *row*). Multiply the first equation by 3 and the second by 2:

$$\begin{aligned} -6x + 9y &= 15 \qquad (R_3 \leftarrow 3R_1) \\ 6x + 18y &= -6 \qquad (R_4 \leftarrow 2R_2) \end{aligned}$$

Note the labelling: the equation $-6x + 9y = 15$ (called R_3) was obtained by multiplying R_1 by 3; we use the same self-explanatory labelling for R_4, and for the remaining calculations. Adding the two equations, we obtain

$$\begin{aligned} 27y &= 9 \qquad (R_5 \leftarrow R_3 + R_4) \\ 3y &= 1 \qquad (R_6 \leftarrow R_5/9) \end{aligned}$$

In this way, we obtain the system

$$\begin{aligned} -2x + 3y &= 5 \qquad (R_1) \\ 3y &= 1 \qquad (R_6) \end{aligned}$$

which is equivalent to the given system.

Now we calculate y from R_6 and then x from R_1: we get $y = 1/3$, and therefore $-2x + 3(1/3) = 5$ and $x = -2$.

The process that we used at the end of the example (calculating y and then x from equations involving only one variable) is called *back substitution*. What made the system of equations R_1 and R_6 *easier* to solve is its special form:

$$\begin{aligned} *x + *y &= * \\ *y &= * \end{aligned}$$

where $*$ represents a real number. The variable y in the second equation could be replaced by x. It does not matter which variable is there; the point is that the second equation involves *one less* variable than the first.

This special form is called *upper-triangular form,* and it allows us to solve the system by back substitution. In the case of a 3×3 system, the upper-triangular form looks like

$$\begin{aligned} *x + *y + *z &= * \\ *y + *z &= * \\ *z &= * \end{aligned}$$

Again, what matters is that (starting from the second row) an equation has one variable less than the equation above it.

Example 5.6 Solving a 3×3 Linear System

Solve the system

$$\begin{aligned} 2x + y - 3z &= -5 && (R_1) \\ 3x - y + 4z &= 12 && (R_2) \\ 4x + 3y + z &= 3 && (R_3) \end{aligned}$$

▶ To start reducing the system to upper-triangular form, we eliminate the x terms from the second and the third equation. Multiply the first equation by 3 and the second by -2 and add them up:

$$\begin{aligned} 6x + 3y - 9z &= -15 && (R_4 \leftarrow 3R_1) \\ -6x + 2y - 8z &= -24 && (R_5 \leftarrow (-2)R_2) \\ 5y - 17z &= -39 && (R_6 \leftarrow R_4 + R_5) \end{aligned}$$

Similarly, we remove the $4x$ term from the third equation:

$$\begin{aligned} -4x - 2y + 6z &= 10 && (R_7 \leftarrow (-2)R_1) \\ 4x + 3y + z &= 3 && (R_3) \\ y + 7z &= 13 && (R_8 \leftarrow R_7 + R_3) \end{aligned}$$

The system

$$\begin{aligned} 2x + y - 3z &= -5 && (R_1) \\ 5y - 17z &= -39 && (R_6) \\ y + 7z &= 13 && (R_8) \end{aligned}$$

is equivalent to the given system. Next, we need to eliminate one of y and z from either R_6 or R_8. We choose to eliminate y from R_8:

$$\begin{aligned} 5y - 17z &= -39 && (R_6) \\ -5y - 35z &= -65 && (R_9 \leftarrow (-5)R_8) \\ -52z &= -104 && (R_{10} \leftarrow R_6 + R_9) \end{aligned}$$

We are done; the linear system that is equivalent to the given system, and is in upper-triangular form, is

$$\begin{aligned} 2x + y - 3z &= -5 && (R_1) \\ 5y - 17z &= -39 && (R_6) \\ -52z &= -104 && (R_{10}) \end{aligned}$$

To find the solutions, we back-substitute: from R_{10} we get $z = 2$. Substituting $z = 2$ into R_6 yields

$$\begin{aligned} 5y - 17(2) &= -39 \\ 5y &= -5 \\ y &= -1 \end{aligned}$$

Finally, using R_1,

$$\begin{aligned} 2x + (-1) - 3(2) &= -5 \\ 2x &= 2 \\ x &= 1 \end{aligned}$$

Thus, the given system has a unique solution: $x = 1$, $y = -1$, and $z = 2$.

In terms of geometry, we showed that the planes given by the equations R_1, R_2, and R_3 intersect at the single point $(1, -1, 2)$; see Figure 5.2. ▲

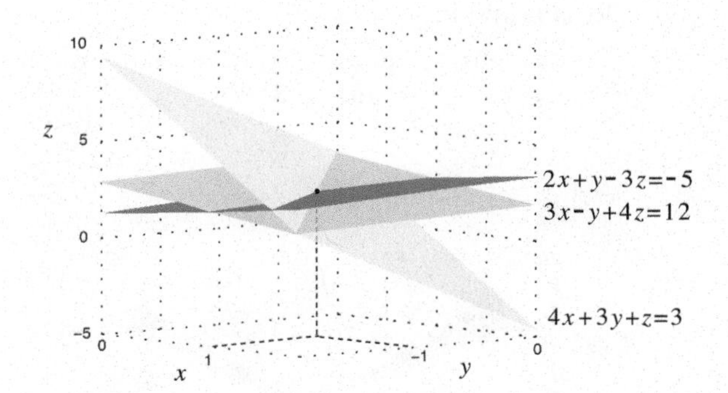

FIGURE 5.2

The intersection of the three planes from Example 5.6

Example 5.7 **Solving a 3 × 3 Linear System**

Solve the system

$$\begin{aligned} 2x - y + 3z &= 12 && (R_1)\\ 4x + 5y + z &= 28 && (R_2)\\ x + 3y - z &= 8 && (R_3) \end{aligned}$$

▶ We start by eliminating the x terms from R_2 and R_3 using the strategy from the previous example. To make it a bit shorter, we combine the steps, as indicated to the right of each equation.

$$\begin{aligned} 2x - y + 3z &= 12 && (R_1)\\ 7y - 5z &= 4 && (R_4 \leftarrow (-2)R_1 + R_2)\\ \frac{7}{2}y - \frac{5}{2}z &= 2 && (R_5 \leftarrow (-1/2)R_1 + R_3) \end{aligned}$$

Next, we work on the last equation:

$$\begin{aligned} 2x - y + 3z &= 12 && (R_1)\\ 7y - 5z &= 4 && (R_4)\\ 0z &= 0 && (R_6 \leftarrow (-1/2)R_4 + R_5) \end{aligned}$$

Equation R_6 holds for any real number z; so let $z = t$, where the parameter t is a real number. From R_4 we get $7y - 5t = 4$ and

$$y = \frac{5}{7}t + \frac{4}{7}$$

Finishing the back substitution, using R_1, we get

$$\begin{aligned} 2x - \left(\frac{5}{7}t + \frac{4}{7}\right) + 3t &= 12\\ 2x &= -3t + \frac{5}{7}t + \frac{4}{7} + 12\\ 2x &= -\frac{16}{7}t + \frac{88}{7}\\ x &= -\frac{8}{7}t + \frac{44}{7} \end{aligned}$$

So, in this case, the linear system has infinitely many solutions, one for each real-number value of t. The solution set, written as

$$\left\{\left(-\frac{8}{7}t + \frac{44}{7}, \frac{5}{7}t + \frac{4}{7}, t\right) \mid t \in \mathbb{R}\right\}$$

or as

$$\begin{bmatrix} x\\ y\\ z \end{bmatrix} = \begin{bmatrix} -8t/7 + 44/7\\ 5t/7 + 4/7\\ t \end{bmatrix} = \begin{bmatrix} 44/7\\ 4/7\\ 0 \end{bmatrix} + t\begin{bmatrix} -8/7\\ 5/7\\ 1 \end{bmatrix} \qquad (5.8)$$

$t \in \mathbb{R}$, is a line in space.

The given system represents three planes in space that intersect along the line (5.8), as shown in Figure 5.3.

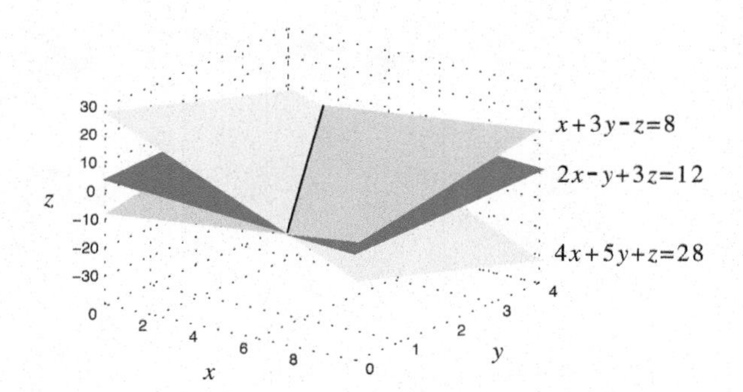

FIGURE 5.3

The planes of Example 5.7 intersect along a line

Summary A **system of linear equations** can have a unique solution, infinitely many solutions, or no solutions at all. When the system has a small number of variables (two or three), we can use **geometry** to reason about its solutions. In general, however, we rely on purely algebraic techniques. We perform operations on equations to reduce the system to **upper-triangular form.** Then we use **back substitution** to find the solutions of the system, one variable at a time.

5 Exercises

1. Find a so that $(3, 4, a)$ is a solution of the linear equation $x - y + 3z = 11$.

2. Find a so that $(x, y, z, w) = (a + 4, a, -5, 1)$ is a solution of the linear equation $2x + 3y - z - w = 12$.

3. Using geometric reasoning, determine the number of solutions of the system

$$\begin{aligned} x + y &= 2 \\ 2x - 2y &= 5 \end{aligned}$$

Confirm your answer by finding the solutions algebraically.

4. Using geometric reasoning, determine the number of solutions of the system

$$\begin{aligned} 3x - y &= 4 \\ 6x - 2y &= 8 \end{aligned}$$

Confirm your answer by finding the solutions algebraically.

5–14 ▪ In each case, describe the solution set geometrically and algebraically. If a solution set contains infinitely many solutions, give the algebraic solution in the form of parametric equations.

5. $4x - 2y = 0$

6. $x + 4y - 9 = 0$

7. $2x - y + z = 4$

8. $3x + 4y + 2z = 5$

9. $\begin{aligned} 3x + y &= 4 \\ x - y &= 0 \end{aligned}$

10. $\begin{aligned} -2x + 2y &= 6 \\ x - y &= -3 \end{aligned}$

11. $\begin{aligned} 10x - 4y &= 2 \\ 5x - 2y &= 1 \end{aligned}$

12. $\begin{aligned} 3x - 4y &= 6 \\ 6x - 8y &= 11 \end{aligned}$

13. $\begin{aligned} 2x - y - z &= 4 \\ 3x + z &= -2 \end{aligned}$

14. $\begin{aligned} 2x - 4y - 8z &= 10 \\ -x + 2y + 4z &= -5 \end{aligned}$

15. Find the solution of the general 2×2 linear system

$$Ax + By = P$$
$$Cx + Dy = Q$$

assuming that $AD - BC \neq 0$.

16. As an exercise in using subscripts, find the solution of the system

$$a_{11}x_1 + a_{12}x_2 = b_1$$
$$a_{21}x_1 + a_{22}x_2 = b_2$$

assuming that $a_{11}a_{22} - a_{12}a_{21} \neq 0$.

17–30 ▪ Reduce each system to upper-triangular form and solve.

17. $2x + 10y = 6$
$-x + y = 9$

18. $3x + y = -12$
$-x - y = 6$

19. $7x - 2y = 2$
$7x - 3y = 2$

20. $-4x + y = 12$
$8x + 2y = -12$

21. $12x + y = 6$
$2x + 7y = 1$

22. $-x - 2y = 2$
$2x + 4y = -4$

23. $3x - y + 4z = 19$
$x + y + 2z = 7$
$-2x - y + z = 0$

24. $x + 2y + z = 7$
$3x - y = -3$
$2x + 3y + 4z = 13$

25. $7x - y - z = 0$
$y + 6z = 37$
$3x + 2y = 5$

26. $8x - 2y - 5z = 16$
$5x - y = 10$
$x + y + z = 2$

27. $x - y + z = 0$
$3x + z = 5$
$5x - 2y + 3z = 5$

28. $4x + y + z = 11$
$3x + y + 3z = 7$
$x - 2z = 4$

29. $x - 2y + z = 4$
$x - y = 8$
$2x - 4y + 2z = 5$

30. $x + y + z = 10$
$3x - 5z = 2$
$4x + y - 4z = 8$

31. Assume that the coefficients in the 2×2 linear system

$$Ax + By = P$$
$$Cx + Dy = Q$$

satisfy $AD - BC = 0$ and $A \neq 0$. Find the conditions on P and Q so that the system has

(a) infinitely many solutions

(b) no solutions

32. Find all values of a (or else say that it does not exist) so that the system

$$ax - y = 2$$
$$6x - 3y = 6$$

has

(a) a unique solution

(b) infinitely many solutions

(c) no solutions

33. Find all values of a (or else say that it does not exist) so that the system

$$\begin{aligned} ax - y &= 2 \\ x - y &= 3 \end{aligned}$$

has

(a) a unique solution

(b) infinitely many solutions

(c) no solutions

34. Show that the homogeneous system

$$\begin{aligned} Ax + By &= 0 \\ Cx + Dy &= 0 \end{aligned}$$

has a unique solution $x = 0, y = 0$ if $AD - BC \neq 0$.

35. Find all solutions of the homogeneous system

$$\begin{aligned} Ax + By &= 0 \\ Cx + Dy &= 0 \end{aligned}$$

if $AD - BC = 0$ and $A \neq 0$.

6 Gaussian Elimination

Motivated by the way we solved the 3×3 system in the last section, we now develop a general method for solving linear systems, called **Gaussian elimination.** To simplify calculations and to keep track of the steps involved, we introduce **matrices.** In this section, we use matrices for solving linear systems only. Our extensive study of matrices, their properties, and their applications starts in Section 8.

Matrices and Linear Systems

We start by defining a matrix and discussing the properties that we will need in this section.

Definition 18 **Matrix**

A *matrix* A is a rectangular table (array) of numbers

$$A = \begin{bmatrix} a_{11} & a_{12} & \cdots & a_{1n} \\ a_{21} & a_{22} & \cdots & a_{2n} \\ \cdots & \cdots & \cdots & \cdots \\ a_{m1} & a_{m2} & \cdots & a_{mn} \end{bmatrix}$$

The real numbers a_{11}, a_{12}, a_{13}, $\ldots$, a_{mn} are called the *entries* of A.

The entries a_{i1}, a_{i2}, $\ldots$, a_{in} form the *ith row* of A, and a_{1j}, a_{2j}, $\ldots$, a_{mj} define the *jth column* of A.

A matrix with m rows and n columns is called an *$m \times n$ matrix.* ▲

A common way of listing entries in an $m \times n$ matrix A is to write

$$[a_{ij}], \text{ where } 1 \leq i \leq m \text{ and } 1 \leq j \leq n$$

Recall that we already used double subscripts in Definition 16 and formula (5.4) in Section 5 when we wrote a linear system in general form. The first subscript represents a row, and the second represents a column. Thus, we can locate the entry a_{ij} by finding the real number located at the intersection of the ith row and the jth column.

Consider the 3×4 matrix

$$A = \begin{bmatrix} 2 & 7 & 6 & -13 \\ 17 & 0 & -4 & 0 \\ 2 & -12 & -5 & 11 \end{bmatrix}$$

To find a_{23}, we look for the entry in the second row and the third column of A: $a_{23} = -4$. Likewise, $a_{33} = -5$, $a_{21} = 17$, $a_{12} = 7$, and so on. The second row of A is the 1×4 matrix (also called a *row vector*)

$$\begin{bmatrix} 17 & 0 & -4 & 0 \end{bmatrix}$$

and the third column of A is the 3×1 matrix (also called a *column vector*)

$$\begin{bmatrix} 6 \\ -4 \\ -5 \end{bmatrix}$$

Thus, we can view a vector as a special matrix that has either one row or one column.

Now we explain how to relate matrices to systems of linear equations.

With an $m \times n$ linear system

$$\begin{aligned} a_{11}x_1 + a_{12}x_2 + \cdots + a_{1n}x_n &= b_1 \\ a_{21}x_1 + a_{22}x_2 + \cdots + a_{2n}x_n &= b_2 \\ &\cdots \\ a_{m1}x_1 + a_{m2}x_2 + \cdots + a_{mn}x_n &= b_m \end{aligned}$$

we associate the $m \times n$ *coefficient matrix* (or the *matrix of the system*)

$$A = \begin{bmatrix} a_{11} & a_{12} & \cdots & a_{1n} \\ a_{21} & a_{22} & \cdots & a_{2n} \\ \cdots & \cdots & \cdots & \cdots \\ a_{m1} & a_{m2} & \cdots & a_{mn} \end{bmatrix}$$

the $m \times 1$ *matrix* (*column vector*) of *constant terms* (*free coefficients*)

$$\mathbf{b} = \begin{bmatrix} b_1 \\ b_2 \\ \cdots \\ b_m \end{bmatrix}$$

and the $m \times (n+1)$ *augmented matrix*

$$[A|\mathbf{b}] = \left[\begin{array}{cccc|c} a_{11} & a_{12} & \cdots & a_{1n} & b_1 \\ a_{21} & a_{22} & \cdots & a_{2n} & b_2 \\ \cdots & \cdots & \cdots & \cdots & \cdots \\ a_{m1} & a_{m2} & \cdots & a_{mn} & b_m \end{array}\right] \tag{6.1}$$

In the augmented matrix we use a vertical line to visually separate the column of constant terms from the coefficients of the variables. If a variable is missing from an equation, we place a zero in the corresponding place in the matrix. For example, if

$$\begin{aligned} 2x + y - 3z &= -5 \\ 3x - 4z &= 12 \\ 3y + z &= 0 \end{aligned}$$

then

$$A = \begin{bmatrix} 2 & 1 & -3 \\ 3 & 0 & -4 \\ 0 & 3 & 1 \end{bmatrix}, \quad \mathbf{b} = \begin{bmatrix} -5 \\ 12 \\ 0 \end{bmatrix}, \quad \text{and} \quad [A|\mathbf{b}] = \left[\begin{array}{ccc|c} 2 & 1 & -3 & -5 \\ 3 & 0 & -4 & 12 \\ 0 & 3 & 1 & 0 \end{array}\right]$$

Why matrices?

Note that in solving a linear system we work with the coefficients and the constant terms in the equations, and the augmented matrix contains all these numbers. So, instead of obtaining an equivalent system by manipulating equations, we manipulate the rows in the augmented matrix.

The operations that we are allowed to perform on an augmented matrix mirror those that we used on equations in Section 5.

Definition 19 Elementary Row Operations

Given a matrix, the following operations are *elementary row operations:*

(a) multiplying or dividing a row by a non-zero number

(b) adding a multiple of a row to another row, or subtracting a multiple of a row from another row

(c) interchanging any two rows

Now we know what operations we are allowed to use to reduce an augmented matrix. But what form do we reduce it to?

In Section 5 we did not formally use augmented matrices. Now that we know what they are, we see that in Example 5.6 the augmented matrix

$$\left[\begin{array}{rrr|r} 2 & 1 & -3 & -5 \\ 3 & -1 & 4 & 12 \\ 4 & 3 & 1 & 3 \end{array}\right]$$

was reduced to the upper-triangular form

$$\left[\begin{array}{rrr|r} 2 & 1 & -3 & -5 \\ 0 & 5 & -17 & -39 \\ 0 & 0 & -52 & -104 \end{array}\right] \tag{6.2}$$

In Example 5.7 the augmented matrix

$$\left[\begin{array}{rrr|r} 2 & -1 & 3 & 12 \\ 4 & 5 & 1 & 28 \\ 1 & 3 & -1 & 8 \end{array}\right]$$

was reduced to

$$\left[\begin{array}{rrr|r} 2 & -1 & 3 & 12 \\ 0 & 7 & -5 & 4 \\ 0 & 0 & 0 & 0 \end{array}\right] \tag{6.3}$$

The reduced forms (6.2) and (6.3) exhibit a "staircase pattern," which allows us to solve the system by back substitution. The upper-triangular form is an example of a staircase pattern (but is not the only one). We now precisely define what we mean by a staircase pattern.

Definition 20 **Row Echelon Form**

The first non-zero entry in a row is called the *leading entry.*

A matrix A is said to be in *row echelon form* if its entries satisfy the following:

(a) All rows that consist of zeros only are placed at the bottom of A.

(b) Comparing rows with leading entries, the leading entry in a row is to the left of the leading entry in the row below it. ▲

From (b) it follows that all entries below a leading entry must be zero. Both matrices (6.2) and (6.3) are in row echelon form, and so are the following matrices:

$$\begin{bmatrix} 3 & 4 & 7 \\ 0 & 2 & 6 \\ 0 & 0 & 0 \end{bmatrix}, \quad \begin{bmatrix} 1 & 2 & 3 \\ 0 & 4 & 5 \\ 0 & 0 & 6 \end{bmatrix}, \quad \begin{bmatrix} 1 & -2 & 3 & 4 \\ 0 & 0 & 14 & 1 \\ 0 & 0 & 0 & 0 \end{bmatrix}$$

and

$$\begin{bmatrix} 3 & -6 & 11 & 4 \\ 0 & 4 & 1 & 0 \\ 0 & 0 & 0 & 9 \end{bmatrix}, \quad \begin{bmatrix} 1 & 0 & 3 & 0 & 9 & 21 \\ 0 & 0 & -2 & 13 & 0 & -1 \\ 0 & 0 & 0 & 0 & 3 & 7 \\ 0 & 0 & 0 & 0 & 0 & 4 \end{bmatrix}$$

The matrix

$$\begin{bmatrix} 0 & 2 & 6 & 4 \\ 5 & 3 & 0 & 5 \\ 0 & 0 & 0 & 7 \end{bmatrix}$$

is not in row echelon form, since (b) from Definition 20 is not satisfied (the leading entry 2 in the first row is not to the left of the leading entry 5 in the second row).

After we interchange the first two rows, we do obtain row echelon form:

$$\begin{bmatrix} 5 & 3 & 0 & 5 \\ 0 & 2 & 6 & 4 \\ 0 & 0 & 0 & 7 \end{bmatrix}$$

The matrix

$$\begin{bmatrix} 1 & 2 & 10 & 4 \\ 0 & 6 & -1 & -3 \\ 0 & 12 & 2 & 0 \end{bmatrix}$$

is not in row echelon form: the leading entry 6 in the second row is not to the left of 12, which is the leading entry in the third row. We can state another reason: the entries below a leading entry have to be zero; this is true for the leading entry 1 in the first row, but not for the leading entry 6 in the second row.

Multiplying the second row by -2 and adding to the third row, we obtain the matrix

$$\begin{bmatrix} 1 & 2 & 10 & 4 \\ 0 & 6 & -1 & -3 \\ 0 & 0 & 4 & 6 \end{bmatrix}$$

which is in row echelon form.

When the augmented matrix of a system is in row echelon form, the system can easily be solved by back substitution. For instance, the system that corresponds to the augmented matrix

$$\left[\begin{array}{cc|c} 2 & 3 & 9 \\ 0 & 4 & 8 \end{array}\right]$$

is given by

$$\begin{aligned} 2x + 3y &= 9 \\ 4y &= 8 \end{aligned}$$

Thus, $y = 8/4 = 2$, and $2x + 3(2) = 9$, i.e., $x = 3/2$.

The system that corresponds to

$$\left[\begin{array}{ccc|c} 1 & -2 & 3 & 4 \\ 0 & 0 & 4 & 1 \\ 0 & 0 & 0 & 0 \end{array}\right]$$

is given by

$$\begin{aligned} x - 2y + 3z &= 4 \\ 4z &= 1 \end{aligned}$$

By back substitution, $z = 1/4$, and so

$$\begin{aligned} x - 2y + 3 \cdot \frac{1}{4} &= 4 \\ x - 2y &= \frac{13}{4} \end{aligned}$$

This equation has infinitely many solutions. Let $y = t$; then $x = 2t + 13/4$, and the solution set of the system is

$$\left\{ \left(2t + \frac{13}{4}, t, \frac{1}{4} \right) \mid t \in \mathbb{R} \right\}$$

The system that corresponds to

$$\left[\begin{array}{ccc|c} 1 & -2 & 3 & 4 \\ 0 & 1 & 4 & 1 \\ 0 & 0 & 0 & 5 \end{array}\right]$$

is inconsistent: the last row represents the equation

$$0x + 0y + 0z = 5$$

which has no solutions.

Gaussian Elimination

We are ready to start solving linear systems.

Algorithm 1 Gaussian Elimination

To solve a system of linear equations:

(a) Create the augmented matrix of the system.

(b) Using elementary row operations (Definition 19), reduce the augmented matrix to row echelon form (Definition 20).

(c) Use back substitution to find the solutions of the system.

Example 6.1 Solving a System of Equations

Solve the system

$$\begin{aligned} x - 2y + 3z &= 5 \\ 3x + y &= -5 \\ -2x + y + z &= 8 \end{aligned}$$

The augmented matrix is

$$\left[\begin{array}{rrr|r} 1 & -2 & 3 & 5 \\ 3 & 1 & 0 & -5 \\ -2 & 1 & 1 & 8 \end{array}\right] \begin{array}{l} (R_1) \\ (R_2) \\ (R_3) \end{array} \tag{6.4}$$

To transform to row echelon form, we work on one column at a time, going from left to right. Our first goal is to introduce zeros in the first column below the leading entry 1. As in manipulating equations, we keep track of the elementary row operations we use.

The matrix (6.4) is equivalent to

$$\left[\begin{array}{rrr|r} 1 & -2 & 3 & 5 \\ 0 & 7 & -9 & -20 \\ 0 & -3 & 7 & 18 \end{array}\right] \begin{array}{l} (R_4 \leftarrow R_1) \\ (R_5 \leftarrow -3R_1 + R_2) \\ (R_6 \leftarrow 2R_1 + R_3) \end{array}$$

Next, we introduce a zero in the second column, below the leading term 7:

$$\left[\begin{array}{rrr|r} 1 & -2 & 3 & 5 \\ 0 & 7 & -9 & -20 \\ 0 & 0 & 22/7 & 66/7 \end{array}\right] \begin{array}{l} (R_7 \leftarrow R_4) \\ (R_8 \leftarrow R_5) \\ (R_9 \leftarrow 3R_5/7 + R_6) \end{array} \tag{6.5}$$

The two non-zero entries in the third row were calculated as follows:

$$\frac{3}{7}(-9) + 7 = -\frac{27}{7} + \frac{49}{7} = \frac{22}{7}$$

$$\frac{3}{7}(-20) + 18 = -\frac{60}{7} + \frac{126}{7} = \frac{66}{7}$$

The augmented matrix (6.5) is in row echelon form. The corresponding equations are

$$\begin{aligned} x - 2y + 3z &= 5 \\ 7y - 9z &= -20 \\ \frac{22}{7}z &= \frac{66}{7} \end{aligned}$$

From the last equation, $z = 3$. By back substitution,

$$7y - 9(3) = -20$$
$$7y = 7$$

and $y = 1$. Finally, the value of x is

$$x - 2(1) + 3(3) = 5$$
$$x = -2$$

Example 6.2 **Reasoning about Solutions**

Suppose that we are solving a 3×3 system and end up with the following augmented matrix in row echelon form:

$$\left[\begin{array}{ccc|c} 1 & 2 & 7 & 3 \\ 0 & 0 & 0 & 4 \\ 0 & 0 & 0 & 0 \end{array}\right]$$

What can we conclude about the solutions?

▶ The last row represents the equation

$$0x + 0y + 0z = 0$$

which is satisfied for all numbers x, y, and z. In other words, this equation gives no restrictions on the values of the variables.

Rewriting the middle row as an equation, we get

$$0x + 0y + 0z = 4$$

which does not hold for any value of x, y, or z. We conclude that the given system has no solutions.

Thus, if the only non-zero entry in a row in an augmented matrix is in the right-most location, then the corresponding linear system has no solutions (in other words, it is an inconsistent linear system).

A linear system that has more variables than equations is called *under-determined.* So an $m \times n$ system is under-determined if $m < n$. If a system has fewer variables than equations, then it is *over-determined.* An $m \times n$ system is over-determined if $m > n$.

Usually, over-determined systems are inconsistent (i.e., have no solutions) and under-determined systems have infinitely many solutions. (However, it does not have to be like that; see Exercises 34 to 36.)

Example 6.3 **Using Gaussian Elimination**

Find all solutions of the 3×2 (i.e., over-determined) linear system

$$-x + y = 0$$
$$2x + 3y = 1$$
$$3x - y = 4$$

▶ The augmented matrix is

$$\left[\begin{array}{cc|c} -1 & 1 & 0 \\ 2 & 3 & 1 \\ 3 & -1 & 4 \end{array}\right] \begin{array}{c} (R_1) \\ (R_2) \\ (R_3) \end{array}$$

First, we eliminate the entries below -1 in the first column:

$$\left[\begin{array}{cc|c} -1 & 1 & 0 \\ 0 & 5 & 1 \\ 0 & 2 & 4 \end{array}\right] \quad \begin{array}{l} (R_4 \leftarrow R_1) \\ (R_5 \leftarrow 2R_1 + R_2) \\ (R_6 \leftarrow 3R_1 + R_3) \end{array}$$

Next, we remove the leading entry 2 in the third row:

$$\left[\begin{array}{cc|c} -1 & 1 & 0 \\ 0 & 5 & 1 \\ 0 & 0 & 18/5 \end{array}\right] \quad \begin{array}{l} (R_7 \leftarrow R_4) \\ (R_8 \leftarrow R_5) \\ (R_9 \leftarrow (-2/5)R_5 + R_6) \end{array}$$

Because of the non-zero entry in the last row $(18/5)$, the system has no solutions (see the comment following Example 6.2).

Our conclusion implies that the three lines $-x + y = 0$, $2x + 3y = 1$, and $3x - y = 4$ do not have a point in common. See Figure 6.1: each pair of lines does have an intersection, but no point lies on all three lines. ▲

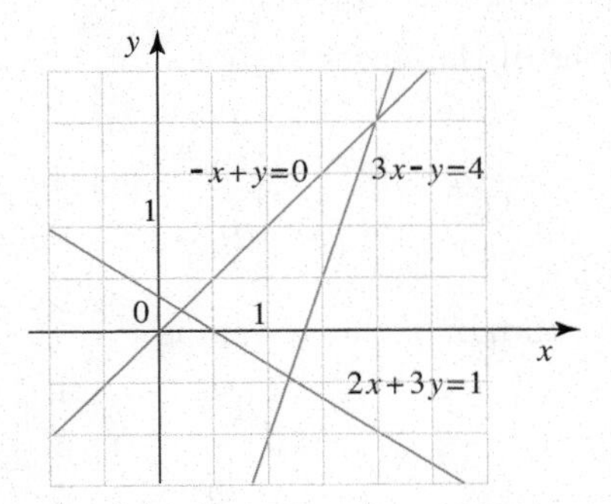

FIGURE 6.1
The lines of Example 6.3

Example 6.4 Using Gaussian Elimination

Find all solutions of the 2×3 (i.e., under-determined) linear system

$$\begin{aligned} -x + 4y - z &= 2 \\ 2x + y - 3z &= 5 \end{aligned}$$

▶ The augmented matrix is

$$\left[\begin{array}{ccc|c} -1 & 4 & -1 & 2 \\ 2 & 1 & -3 & 5 \end{array}\right]$$

To reduce to row echelon form we need to introduce a zero in the place of the 2 in the second row. Multiplying the first row by 2 and adding to the second row, we obtain the reduced matrix

$$\left[\begin{array}{ccc|c} -1 & 4 & -1 & 2 \\ 0 & 9 & -5 & 9 \end{array}\right]$$

with the corresponding system

$$\begin{aligned} -x + 4y - z &= 2 \\ 9y - 5z &= 9 \end{aligned}$$

To solve $9y - 5z = 9$ (one equation, two variables) we designate one variable as a parameter, say, $z = t$. Then

$$\begin{aligned} 9y - 5t &= 9 \\ 9y &= 5t + 9 \\ y &= \frac{5}{9}t + 1 \end{aligned}$$

Using the equation $-x + 4y - z = 2$, we compute x:

$$-x + 4\left(\frac{5}{9}t + 1\right) - t = 2$$
$$-x + \frac{20}{9}t + 4 - t = 2$$
$$x = \frac{11}{9}t + 2$$

The solution of the system is the line

$$\left\{\left(\frac{11}{9}t + 2, \frac{5}{9}t + 1, t\right) \mid t \in \mathbb{R}\right\}$$

obtained as the intersection of the two planes defined by the given system of equations. ▲

Example 6.5 **Solving a System of Linear Equations**

Find all solutions of the linear system

$$\begin{aligned} x + y - 3z + w &= 4 \\ y + 4w &= 1 \\ x + 2y - 3z + 5w &= 5 \end{aligned}$$

▶ The augmented matrix of the system is

$$\left[\begin{array}{cccc|c} 1 & 1 & -3 & 1 & 4 \\ 0 & 1 & 0 & 4 & 1 \\ 1 & 2 & -3 & 5 & 5 \end{array}\right]$$

First, we remove the leading entry 1 in the third row by subtracting the first row from the third row:

$$\left[\begin{array}{cccc|c} 1 & 1 & -3 & 1 & 4 \\ 0 & 1 & 0 & 4 & 1 \\ 0 & 1 & 0 & 4 & 1 \end{array}\right]$$

To finish, subtract the second row from the third row:

$$\left[\begin{array}{cccc|c} 1 & 1 & -3 & 1 & 4 \\ 0 & 1 & 0 & 4 & 1 \\ 0 & 0 & 0 & 0 & 0 \end{array}\right]$$

This row echelon form implies the following system:

$$\begin{aligned} x + y - 3z + w &= 4 \\ y + 4w &= 1 \end{aligned} \tag{6.6}$$

(Compare with the starting system; the row reduction process removed the redundant third equation.) We now use back substitution.

Introducing a parameter, we write $w = t$, and thus $y = 1 - 4t$. The first equation in (6.6) implies

$$x + (1 - 4t) - 3z + t = 4$$
$$x - 3z = 3 + 3t$$

To solve this equation (in two variables) we need another parameter. Let $z = s$; then

$$x = 3 + 3t + 3z = 3 + 3t + 3s$$

and the solution set of the system is given by

$$\{(3 + 3t + 3s, 1 - 4t, s, t) \mid s, t \in \mathbb{R}\}$$

▲

Example 6.6 Lines in Space

Determine whether or not the lines

$$\begin{bmatrix} x \\ y \\ z \end{bmatrix} = \begin{bmatrix} 1 \\ 2 \\ 0 \end{bmatrix} + s \begin{bmatrix} -1 \\ 2 \\ 4 \end{bmatrix} \quad \text{and} \quad \begin{bmatrix} x \\ y \\ z \end{bmatrix} = \begin{bmatrix} 0 \\ 3 \\ -1 \end{bmatrix} + t \begin{bmatrix} 1 \\ 1 \\ 1 \end{bmatrix}$$

intersect.

▶ We are looking for a point that belongs to both lines; i.e., we need to figure out if there is a value of t and a value of s such that

$$\begin{bmatrix} 1 \\ 2 \\ 0 \end{bmatrix} + s \begin{bmatrix} -1 \\ 2 \\ 4 \end{bmatrix} = \begin{bmatrix} 0 \\ 3 \\ -1 \end{bmatrix} + t \begin{bmatrix} 1 \\ 1 \\ 1 \end{bmatrix}$$

Writing this vector equation in coordinates, we obtain

$$\begin{aligned} 1 - s &= t \\ 2 + 2s &= 3 + t \\ 4s &= -1 + t \end{aligned}$$

and, after simplifying,

$$\begin{aligned} s + t &= 1 \\ -2s + t &= -1 \\ -4s + t &= 1 \end{aligned} \tag{6.7}$$

Since the system looks fairly simple, we manipulate equations instead of using the augmented matrix approach.

Multiplying the first equation in (6.7) by 2 and adding to the second equation, we get $3t = 1$, and $t = 1/3$. Multiplying the first equation by 4 and adding to the third equation yields $5t = 5$, i.e., $t = 1$. Clearly, t cannot be equal to two different values. We conclude that the system has no solution, and so the two lines do not intersect. ▲

Summary A system of linear equations can be expressed in the form of an **augmented matrix,** which contains the coefficients of all variables and the constant terms. Using **elementary row operations,** the augmented matrix is reduced to **row echelon form,** which enables us to solve the system by **back substitution,** one variable at a time. This approach to solving a linear system of equations is known as **Gaussian elimination.**

6 Exercises

1. Give an example of a 3×2 linear system that has infinitely many solutions.

2. Is it possible to find a 2×3 linear system that has a unique solution?

3–6 ▪ Give a reason why each matrix is not in row echelon form. By performing one row operation, transform it into row echelon form.

3. $\begin{bmatrix} 1 & 1 \\ 1 & 0 \end{bmatrix}$

4. $\begin{bmatrix} 0 & 0 \\ 1 & 0 \end{bmatrix}$

5. $\begin{bmatrix} 1 & 2 & 3 \\ 0 & 0 & 1 \\ 0 & 0 & 4 \end{bmatrix}$

6. $\begin{bmatrix} 1 & 0 & 3 \\ 0 & 1 & 1 \\ 0 & 1 & 4 \end{bmatrix}$

7–14 ▪ Reduce each matrix to row echelon form.

7. $\begin{bmatrix} 1 & 2 \\ 3 & 4 \end{bmatrix}$

8. $\begin{bmatrix} 0 & 0 & 1 \\ 0 & 2 & 0 \\ 3 & 0 & 0 \end{bmatrix}$

9. $\begin{bmatrix} 1 & 0 & -3 \\ 0 & 4 & 0 \\ -3 & 0 & 1 \end{bmatrix}$

10. $\begin{bmatrix} 1 & 2 \\ 3 & 1 \\ 0 & 4 \end{bmatrix}$

11. $\begin{bmatrix} -2 & 1 & 1 \\ 8 & 0 & 0 \end{bmatrix}$

12. $\begin{bmatrix} 1 & 2 & 3 \\ 1 & 2 & 3 \\ 1 & 2 & 4 \end{bmatrix}$

13. $\begin{bmatrix} 1 & 3 & -2 \\ 2 & 4 & 0 \\ 0 & 4 & 1 \end{bmatrix}$

14. $\begin{bmatrix} -1 & 0 & 1 \\ 7 & 1 & 2 \\ -2 & 1 & 3 \end{bmatrix}$

15. Describe the set of points that belong to the two planes $x - 3y + z = 2$ and $2x + 3y - z = 5$.

16. Find the intersection of the three planes $3x - y - z = 2$, $x + 4z = -1$, and $4x - y + 3z = 1$.

17–22 ▪ Given is the augmented matrix of a linear system. In each case, without formally solving the system, determine the number of solutions.

17. $\left[\begin{array}{ccc|c} 1 & 3 & 0 & 2 \\ 0 & 2 & 4 & 4 \\ 0 & 0 & 0 & 0 \end{array}\right]$

18. $\left[\begin{array}{ccc|c} 1 & 3 & 0 & 2 \\ 0 & 2 & 4 & 4 \\ 0 & 0 & 0 & 6 \end{array}\right]$

19. $\left[\begin{array}{cccc|c} 1 & 1 & -3 & 2 & 3 \\ 0 & 0 & 2 & 4 & 2 \\ 0 & 0 & 0 & 1 & 5 \end{array}\right]$

20. $\left[\begin{array}{cc|c} 1 & 2 & -2 \\ 0 & 1 & 7 \\ 0 & 0 & 0 \end{array}\right]$

21. $\left[\begin{array}{ccc|c} 1 & 3 & 0 & 2 \\ 0 & 2 & 4 & 4 \\ 0 & 0 & 2 & 2 \end{array}\right]$

22. $\left[\begin{array}{ccc|c} 1 & 2 & 3 & 2 \\ 0 & 0 & 0 & 0 \\ 0 & 0 & 0 & 0 \end{array}\right]$

23. Express the vector $[8, -4, 8]$ as a linear combination of the vectors $[3, 0, 1]$, $[1, 2, -1]$, and $[4, 1, 2]$.

24. Express the vector $[13, 22, 7]$ as a linear combination of the vectors $[0, 1, 2]$, $[2, 0, -3]$, and $[3, 4, 0]$.

25–32 ▪ Solve each system using Gaussian elimination.

25. $\begin{aligned} 3x - y + 4z &= 19 \\ x + y + 2z &= 7 \\ -2x - y + z &= 0 \end{aligned}$

26. $\begin{aligned} x + 2y + z &= 7 \\ 3x - y &= -3 \\ 2x + 3y + 4z &= 13 \end{aligned}$

27. $\begin{aligned} x + y - 2z &= 4 \\ 2x + z &= 0 \\ 3x + y - z &= 4 \end{aligned}$

28. $\begin{aligned} x - z &= 1 \\ -2x + 2y + z &= -1 \\ y - z &= 2 \end{aligned}$

29. $\begin{aligned} 7x - y - z &= 0 \\ y + 6z &= 37 \\ 3x + 2y &= 5 \end{aligned}$

30. $\begin{aligned} 8x - 2y - 5z &= 16 \\ 5x - y &= 10 \\ x + y + z &= 2 \end{aligned}$

31. $\begin{aligned} x - y + z &= 0 \\ 3x + z &= 5 \\ 5x - 2y + 3z &= 5 \end{aligned}$

32. $\begin{aligned} 4x + y + z &= 11 \\ 3x + y + 3z &= 7 \\ x - 2z &= 4 \end{aligned}$

33. Using row reduction, solve the system

$$\begin{aligned} Ax + By &= P \\ Cx + Dy &= Q \end{aligned}$$

under the assumptions $AD - BC \neq 0$ and $A \neq 0$.

34. By reducing to row echelon form, show that the under-determined system

$$\begin{aligned} 3x - y + 4z + t &= 4 \\ x + z + 2t &= 6 \\ 4x - y + 5z + 3t &= 7 \end{aligned}$$

has no solutions.

35. Show that the over-determined system

$$\begin{aligned} 3x + y &= 5 \\ 4x + 3y &= 5 \\ 5x - y &= 11 \end{aligned}$$

has a unique solution.

36. Show that the over-determined system

$$\begin{aligned} 2x - y + z &= 6 \\ 2y - z &= 3 \\ 2x + y &= 9 \\ 4y - 2z &= 6 \end{aligned}$$

has infinitely many solutions.

37. Determine whether or not the lines

$$\begin{bmatrix} x \\ y \\ z \end{bmatrix} = \begin{bmatrix} 6 \\ -2 \\ 5 \end{bmatrix} + t \begin{bmatrix} -1 \\ 1 \\ -1 \end{bmatrix} \quad \text{and} \quad \begin{bmatrix} x \\ y \\ z \end{bmatrix} = \begin{bmatrix} 0 \\ 2 \\ 1 \end{bmatrix} + s \begin{bmatrix} 3 \\ -2 \\ 2 \end{bmatrix}$$

$(s, t \in \mathbb{R})$ intersect.

38. Determine whether or not the lines

$$\begin{bmatrix} x \\ y \\ z \end{bmatrix} = \begin{bmatrix} -2 \\ 4 \\ 5 \end{bmatrix} + t \begin{bmatrix} -1 \\ 0 \\ 1 \end{bmatrix} \quad \text{and} \quad \begin{bmatrix} x \\ y \\ z \end{bmatrix} = \begin{bmatrix} 0 \\ 2 \\ 0 \end{bmatrix} + s \begin{bmatrix} -1 \\ 1 \\ 0 \end{bmatrix}$$

$(s, t \in \mathbb{R})$ intersect.

7 Linear Systems in Medical Imaging

The purpose of an imaging technique (X-ray, CT, MRI, ultrasound) is to "see" what is inside a human body, or some other object without physically opening it. We will see how systems of **linear equations** play an essential role in obtaining an image using a computed tomography (CT) scan.

Introduction

To obtain an X-ray image of a human liver (or any other organ), a source of X-ray beams and a photographic film are placed on opposite sides of the liver. The passing X-ray beams produce a *projection* of the liver on the film. This projection is what we see and use for diagnostic and treatment decision purposes.

It can be proven mathematically that the full internal structure of a liver (or another organ, or almost any object) can be obtained by taking a sufficiently large number of projections. This is the principle behind *computed tomography* (CT): based on data collected from multiple projections, we *calculate* the internal structure. In the case of a CT scan, it is the internal structure of a *slice* (a cross-section) of a human body.

First, we illustrate the multiple-projections principle in an example.

Example 7.1 Multiple Projections

A wooden box has four square compartments of equal size, arranged in two rows and two columns (Figure 7.1). An unknown number of iron balls is placed in each compartment. The box is closed and we cannot see the balls.

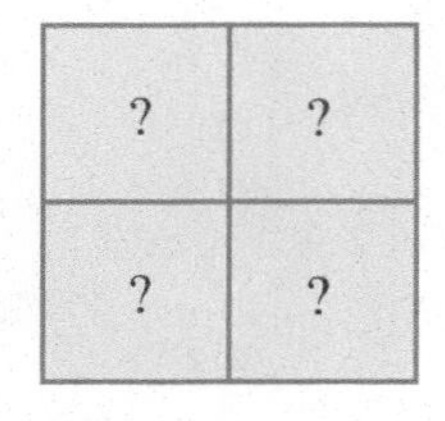

FIGURE 7.1
Box with an unknown number of balls

Our objective is to find the number of balls in each compartment.

We have a scanning device (such as an X-ray machine) that we can position any way we wish around the box. The X-rays cannot detect the walls of the box, but they can detect the iron balls.

To start, we aim the X-rays horizontally and place the detector on the opposite side (Figure 7.2a). Because the X-rays ignore the wooden walls of the box, the detector can only report the total number of balls that the X-ray beam encounters. Assume that the horizontal scan shows eight balls (for the first row) and seven balls (for the second row); see Figure 7.2b. Thus, we know that the two compartments in the first row contain a total of eight balls, and the two compartments in the second row contain a total of seven balls. Clearly, we do not have enough information to figure out how many balls there are in each compartment.

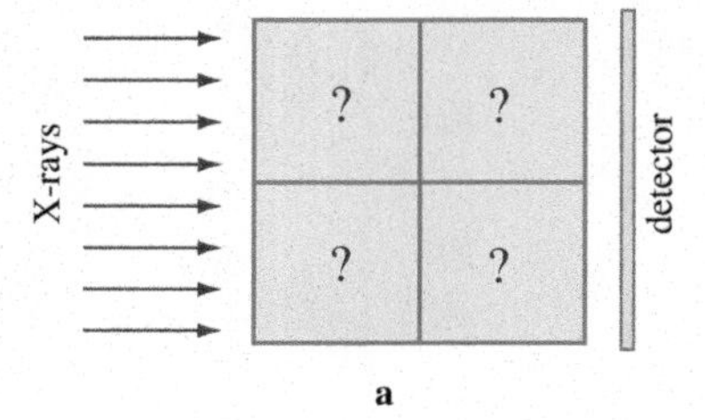

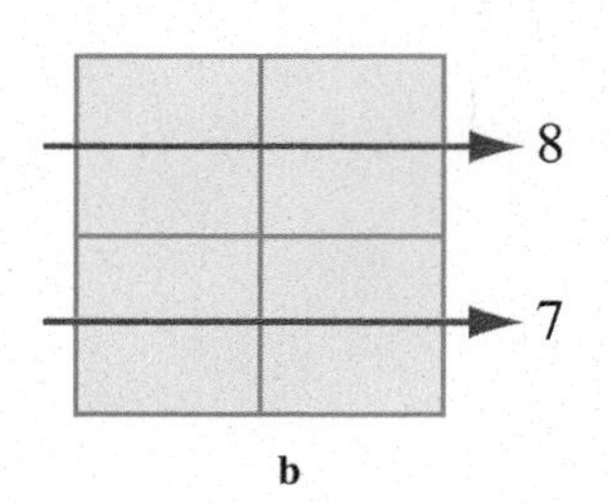

FIGURE 7.2
Horizontal scan

We can move the scanner around, so we rotate it and shoot a set of parallel X-ray beams from below (Figure 7.3a). This time, the detector returns the number 4 for the left column and the number 11 for the right column (Figure 7.3b). Can we figure out now how many balls there are in each compartment?

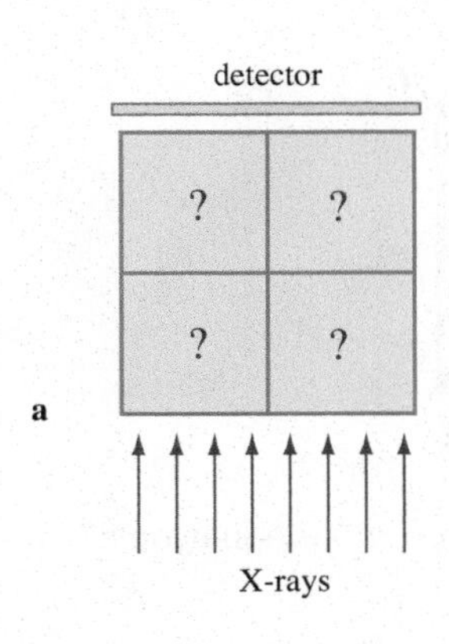

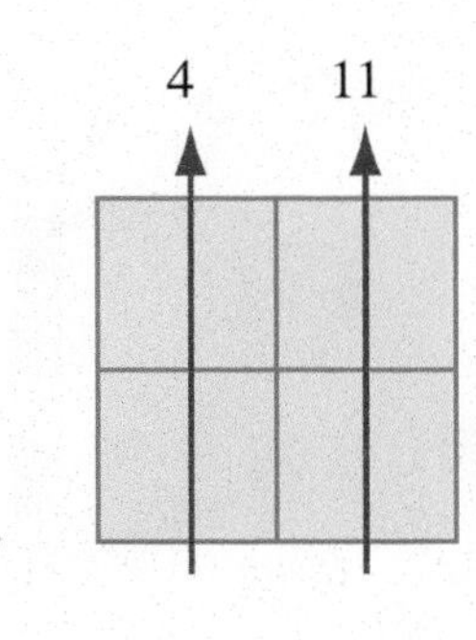

FIGURE 7.3
Vertical scan

Denote the number of balls in the four compartments by x, y, z, and w, respectively (Figure 7.4). Based on the data obtained from the two scans, we obtain a linear system of four equations with four variables:

$$\begin{aligned} x + y &= 8 \\ z + w &= 7 \\ x + z &= 4 \\ y + w &= 11 \end{aligned} \tag{7.1}$$

x	y
z	w

FIGURE 7.4
The four compartments

If we can solve this system, we are done. Let's try row reduction. Form the augmented matrix

$$\left[\begin{array}{cccc|c} 1 & 1 & 0 & 0 & 8 \\ 0 & 0 & 1 & 1 & 7 \\ 1 & 0 & 1 & 0 & 4 \\ 0 & 1 & 0 & 1 & 11 \end{array}\right]$$

Replace the second row with the sum of the first row and the second row, and replace the fourth row with the sum of the third row and the fourth row:

$$\left[\begin{array}{cccc|c} 1 & 1 & 0 & 0 & 8 \\ 1 & 1 & 1 & 1 & 15 \\ 1 & 0 & 1 & 0 & 4 \\ 1 & 1 & 1 & 1 & 15 \end{array}\right]$$

Subtract the second row from the fourth row, and insert the result in the fourth row:

$$\left[\begin{array}{cccc|c} 1 & 1 & 0 & 0 & 8 \\ 1 & 1 & 1 & 1 & 15 \\ 1 & 0 & 1 & 0 & 4 \\ 0 & 0 & 0 & 0 & 0 \end{array}\right]$$

We do not need to go any further. The zeros in the last row indicate that the system has infinitely many solutions, that is, if the variables x, y, z, and w are real numbers. In our case, the variables are non-negative integers, but nevertheless we do not have a unique solution: it's easy to check that $x = 1$, $y = 7$, $z = 3$, and $w = 4$; $x = 2$, $y = 6$, $z = 2$, and $w = 5$; and $x = 0$, $y = 8$, $z = 4$, and $w = 3$ are solutions of (7.1), and there are more.

Hence an important conclusion: in situations such as this one, we need an *over-determined* system (more equations than variables); this is exactly what happens in the case of a CT scan (we'll get to this later).

Assume that we can scan in such a way that we can detect the two diagonals and that the scanner returns the values 9 and 6; see Figure 7.5.

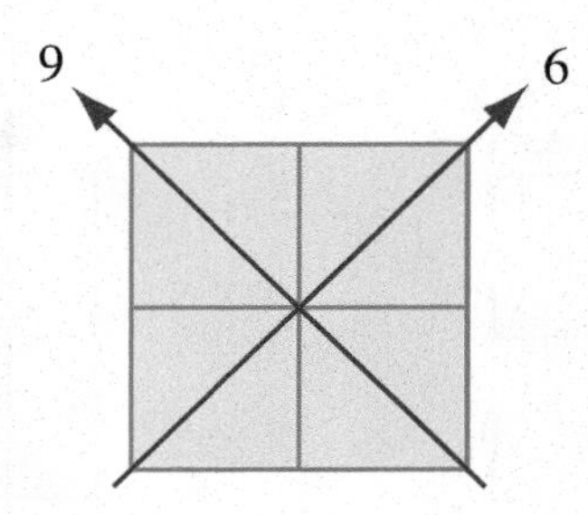

FIGURE 7.5
Diagonal scans

In this way, we obtain two more equations:

$$\begin{aligned} x + w &= 9 \\ y + z &= 6 \end{aligned} \tag{7.2}$$

Merging (7.1) and (7.2), we obtain a system of six equations in four variables, which does have a unique solution. The solution is simple, so we will not use matrices: combining $x + y = 8$ and $x + w = 9$, we obtain

$$8 - y = 9 - w$$
$$w - y = 1$$

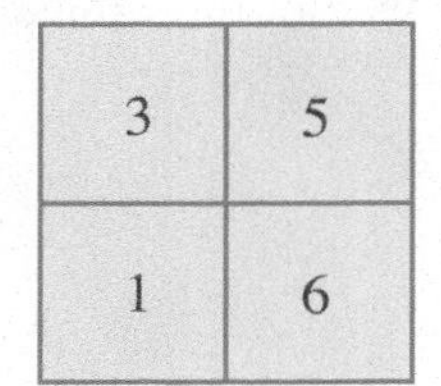

FIGURE 7.6
Solution to the problem in Example 7.1

Adding this equation to $y + w = 11$ yields $2w = 12$ and $w = 6$. Thus, $y = 5$, and from $x + y = 8$ it follows that $x = 3$. Finally, from $x + z = 4$ we get $z = 1$. The solution is shown in Figure 7.6.

Computed Tomography

The purpose of a CT scan is to obtain a sequence of images—parallel cross-sections of an organ (such as a liver or a human brain) taken at predefined distances from each other (see Figure 7.7a).

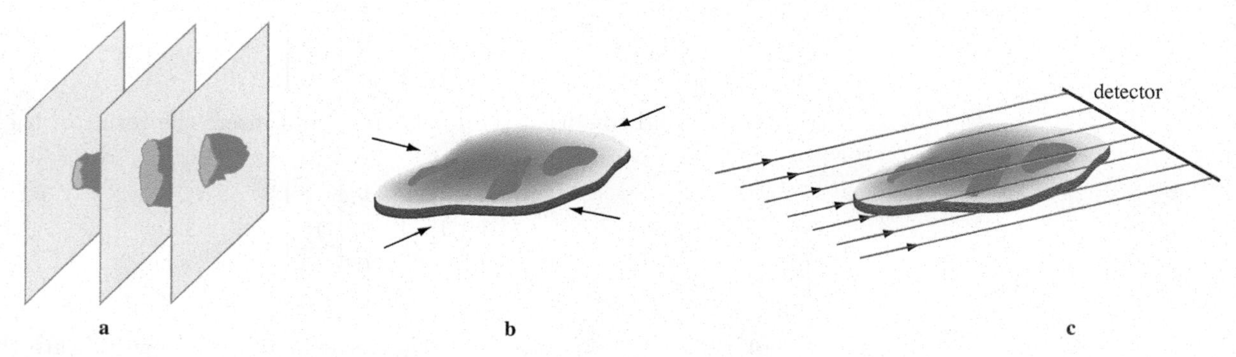

FIGURE 7.7
Scanning to obtain parallel cross-sections; scanning a single tomographic slice

Each cross-section (also called a *tomographic slice*) is examined from multiple directions with a narrow X-ray beam (Figure 7.7b). The detector on the opposite side of the entering beam measures the attenuated (weakened) radiation that remains after the X-rays pass through the slice (Figure 7.7c).

This information about the attenuated radiation is used to calculate the two-dimensional image of a tomographic slice and to produce an image such as the one in Figure 7.8. Let's explain how this is done.

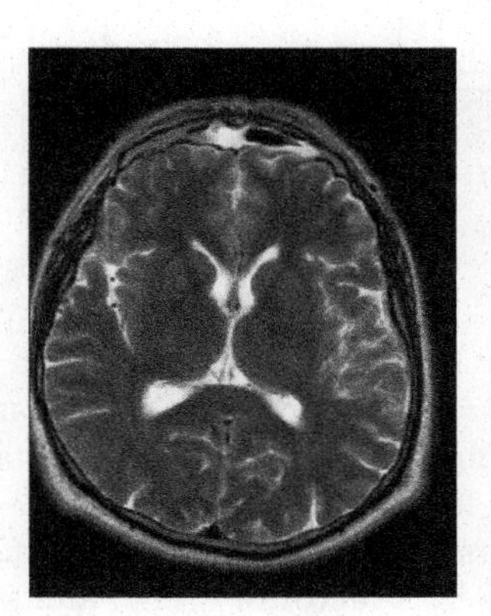

FIGURE 7.8

Tomographic slice of a brain

(Courtesy of Miroslav Lovrić.)

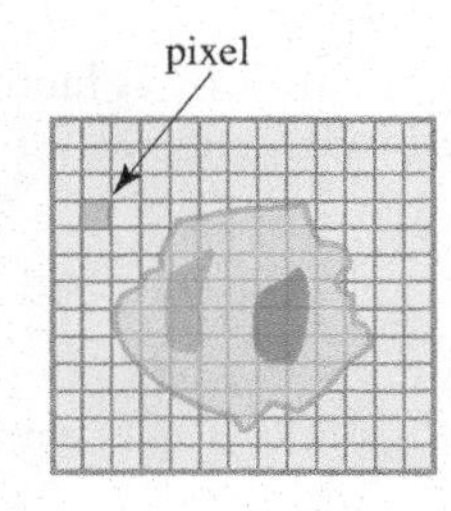

FIGURE 7.9

Placing a grid over a tomographic slice

A square grid is placed over a tomographic slice; see Figure 7.9. A unit square in the grid is called a *pixel.* The size of the grid differs depending on the desired resolution and on one extra fact: the finer the scanning resolution, the longer the exposure to the harmful X-rays (thus, it's important to achieve a good balance between the resolution and the exposure). The first CT scans used an 80×80 square grid, with each pixel measuring 3 mm $\times$ 3 mm. Today, grids with a pixel size below 1 mm $\times$ 1 mm are routinely used.

The final objective of the mathematical calculations is to assign a number to each pixel in the grid. This number characterizes the material that occupies the pixel; in the case of a brain the material could be part of a bone, white matter, grey matter, a blood vessel, a blood clot, cerebrospinal fluid, a tumour, a cyst, and so on.

The number assigned to a pixel depends on the degree to which the material within that pixel attenuates (weakens) the X-ray beam passing through it. This works because it is known that different materials (bone, tissue, blood, water, etc.) attenuate an X-ray beam by different rates.

We assume that a pixel size is $d \times d$ and start with the case where the whole grid consists of a single pixel (Figure 7.10a).

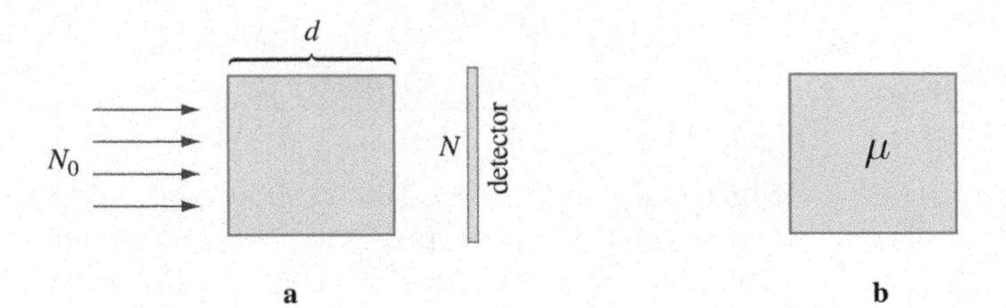

FIGURE 7.10

One-pixel grid

The intensity, N_0, of the incoming radiation is known (could be measured, for instance, in the number of photons), and the detector on the opposite side measures the attenuated intensity, N. The intensity decays exponentially:

$$N = N_0 e^{-\mu d}$$

where $\mu > 0$. Thus,

$$\begin{aligned} e^{-\mu d} &= \frac{N}{N_0} \\ -\mu d &= \ln(N/N_0) \\ \mu &= -\frac{1}{d}\ln(N/N_0) \end{aligned}$$

Since N, N_0, and d are known, we can calculate μ. (Note that $N < N_0$, so the logarithm term is negative, which implies that $\mu > 0$.)

The constant μ is the number that we assign to the pixel (Figure 7.10b).

Next, assume that we have three pixels in a row, as shown in Figure 7.11.

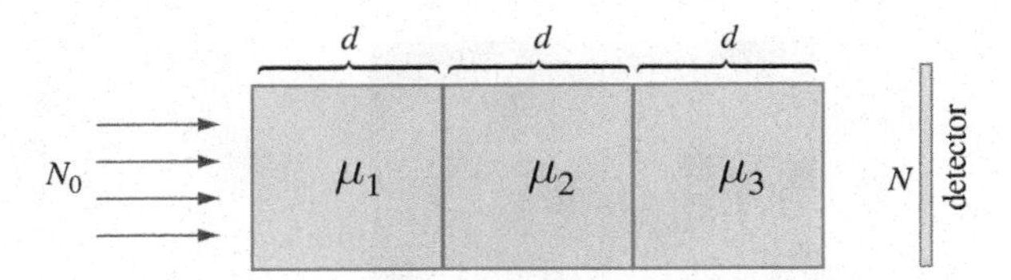

FIGURE 7.11
Grid with three horizontally placed pixels

This time, the X-ray beam has to pass through three pixels, and we need to figure out μ_1, μ_2, and μ_3, which characterize the material in each pixel. The incoming intensity N_0 is attenuated by the factor $e^{-\mu_1 d}$ at the moment when it reaches the border between the left and the middle pixel. Thus, $N_0 e^{-\mu_1 d}$ is the intensity of the beam as it enters the middle pixel. The material in the middle pixel further attenuates the beam by a factor of $e^{-\mu_2 d}$, and so the intensity of the beam leaving the middle pixel is

$$N_0 e^{-\mu_1 d} e^{-\mu_2 d} = N_0 e^{-\mu_1 d - \mu_2 d}$$

After passing through the right pixel, the beam is further attenuated by a factor of $e^{-\mu_3 d}$. The intensity of the beam coming out (and measured by the detector) is

$$\begin{aligned} N &= N_0 e^{-\mu_1 d - \mu_2 d} e^{-\mu_3 d} \\ &= N_0 e^{-(\mu_1 + \mu_2 + \mu_3) d} \end{aligned}$$

Solving as before, we get

$$\mu_1 + \mu_2 + \mu_3 = -\frac{1}{d} \ln(N/N_0)$$

So, we do not know the values of each of μ_1, μ_2, and μ_3, but we know their sum. (Does this ring a bell? The balls in the box?)

Assume that we work with a 100×100 grid. Then each row in a grid (Figure 7.12) yields an equation in 100 variables of the form

$$\mu_1 + \mu_2 + \cdots + \mu_{100} = -\frac{1}{d} \ln(N/N_0)$$

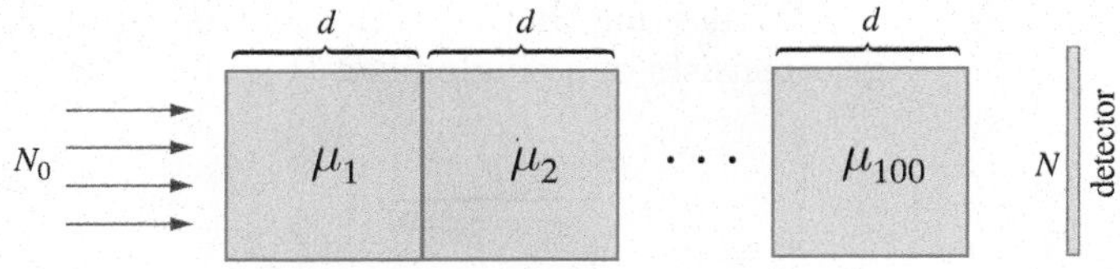

FIGURE 7.12
One row in a 100×100 grid

Since the grid is of size 100×100, it contains a total of ten thousand pixels. To find the μ value for each pixel, we need many equations, certainly more than 10,000 (in Example 7.1 we realized that the linear system we obtain in this way needs to be over-determined in order to have a *unique* solution).

Scanning in one direction (say, horizontal) produces 100 equations (it was two in the case of a 2×2 grid; see Figure 7.2a). Scanning in the perpendicular direction produces another set of 100 equations; that's 200, and we need more than 10,000. Clearly, we need to scan from many directions.

The actual CT scanner is built so that it can rotate, with the object being scanned placed at or near the centre of rotation. Assuming that we perform one scan for each degree of rotation (so 180 scans in total), and assuming that each position generates 100 equations, we obtain 18,000 equations.

So, what is really needed is a good algorithm that can solve this system of 18,000 equations in 10,000 variables correctly and quickly. Once this is accomplished, each pixel ends up with its μ value. The last step consists of a numeric conversion: the μ value is converted to a code representing a shade of grey, or a colour, so that the grid is shown as a greyscale (as in Figure 7.8) or a colour picture. In Figure 7.13 we see individual pixels coloured in shades of grey.

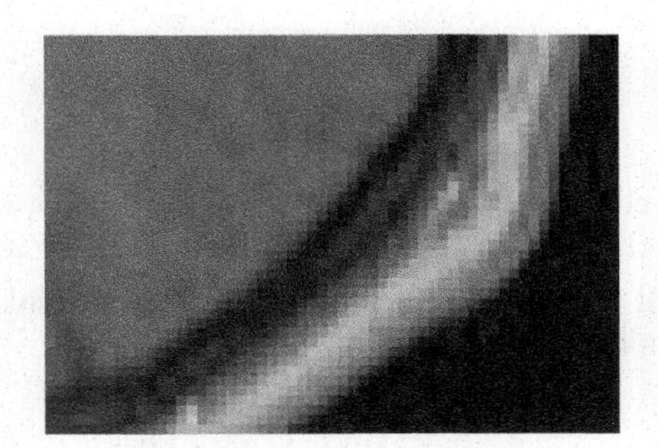

FIGURE 7.13
Magnified image showing greyscale pixels
(Courtesy of Miroslav Lovrić.)

By manipulating the conversion algorithm between the μ values and the shades of grey (or colour), various features of the scan can be enhanced and/or seen better in the image.

We do not go into detail of how this large system of linear equations is solved. Some algorithms that are used in practice are indeed based on elementary linear algebra (using orthogonal projection to obtain an approximate solution), whereas some use tools from other areas of mathematics (such as Fourier transforms).

One more comment: when the scanner is positioned at an angle (rather than parallel to the sides of the grid), the attenuation equations need to be adjusted, since the X-ray beams no longer travel the same distance through each pixel.

For instance, if we scan at 45°, the beams b_1, b_2, and b_3 in Figure 7.14 travel different distances through the pixels.

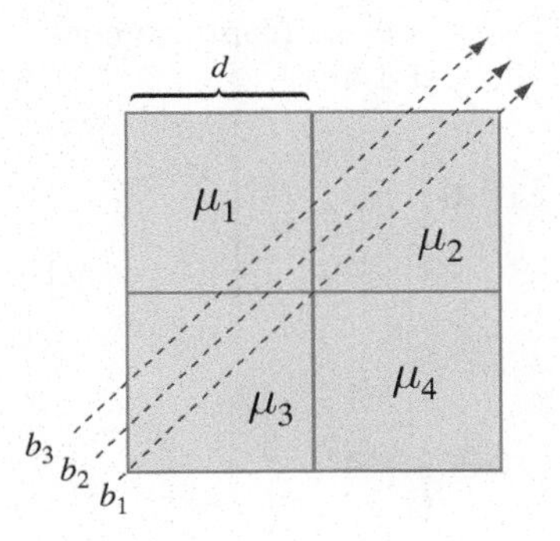

FIGURE 7.14
Scanning at 45°

For instance, the equation for beam b_1 is (we use $\exp(x)$ instead of e^x; otherwise the symbols in the exponent would be too small)

$$N = N_0 \exp\left(-\mu_3 d\sqrt{2} - \mu_2 d\sqrt{2}\right)$$

Assuming that b_2 enters the pixel at a point $d/4$ units away from the corner, its attenuation equation reads

$$N = N_0 \exp\left(-\mu_3 d\frac{3\sqrt{2}}{4} - \mu_1 d\frac{\sqrt{2}}{4} - \mu_2 d\frac{3\sqrt{2}}{4}\right)$$

(Pythagorean theorem). A similar equation is obtained for the beam b_3. There are other ways to adjust the attenuation equations, but we do not discuss them here.

Summary

In order to create an image generated by a CT scan, a large system of linear equations needs to be solved. The main principle behind the equations is that the intensity of the X-ray radiation attenuates (decays) as it passes through a slice of the human body.

8 Matrices

In this section, we learn about various **matrix operations** and their properties. At the end, we define the **determinant** of a matrix.

Recall that an $m \times n$ matrix A is a rectangular table of real numbers arranged in m rows and n columns:

$$A = \begin{bmatrix} a_{11} & a_{12} & \cdots & a_{1n} \\ a_{21} & a_{22} & \cdots & a_{2n} \\ \cdots & \cdots & \cdots & \cdots \\ a_{m1} & a_{m2} & \cdots & a_{mn} \end{bmatrix} \tag{8.1}$$

Instead of displaying elements in table format, we often use the abbreviated notation $A = [a_{ij}]$, where $i = 1, 2, \ldots, m$ and $j = 1, 2, \ldots, n$. Matrices are denoted in uppercase letters, and their entries using the corresponding lowercase letter.

A matrix all of whose entries are zero is called a *zero matrix* and is denoted by 0.

A matrix $A = [a_{ij}]$ of type $n \times n$, i.e., with the same number of rows and columns, is called a *square matrix* (or a *square matrix of size* n). The entries $a_{11}, a_{22}, \ldots, a_{nn}$ form the *main diagonal* of A; all other entries are referred to as the *off-diagonal entries.*

The square matrix of size n all of whose diagonal entries are 1 and all of whose off-diagonal entries are 0 is called the *identity matrix* (*of size* n), and is denoted by I_n (or just I, if its size is clear from the context, or is not relevant). The identity matrices of size 2, 3, and 4 are

$$I_2 = \begin{bmatrix} 1 & 0 \\ 0 & 1 \end{bmatrix}, \quad I_3 = \begin{bmatrix} 1 & 0 & 0 \\ 0 & 1 & 0 \\ 0 & 0 & 1 \end{bmatrix}, \quad I_4 = \begin{bmatrix} 1 & 0 & 0 & 0 \\ 0 & 1 & 0 & 0 \\ 0 & 0 & 1 & 0 \\ 0 & 0 & 0 & 1 \end{bmatrix}$$

Basic Matrix Operations

We define several algebraic operations on matrices.

Definition 21 **Equal Matrices**

Two matrices $A = [a_{ij}]$ and $B = [b_{ij}]$ are said to be *equal,* denoted by $A = B$, if they are of the same size (say, $m \times n$) and the corresponding entries are equal:

$$a_{ij} = b_{ij}$$

for $i = 1, 2, \ldots, m$ and $j = 1, 2, \ldots, n$. ▲

Thus, if

$$\begin{bmatrix} a & b \\ c & d \end{bmatrix} = \begin{bmatrix} 0 & 4 \\ 2 & 3 \end{bmatrix}$$

then $a = 0$, $b = 4$, $c = 2$, and $d = 3$. Although the matrices

$$\begin{bmatrix} 2 \\ 3 \\ 1 \end{bmatrix} \quad \text{and} \quad \begin{bmatrix} 2 & 3 & 1 \end{bmatrix}$$

contain the same entries, 2, 3, and 1, they are not equal because they are not of the same size.

Definition 22 **Addition, Subtraction, and Multiplication by a Scalar**

Assume that $A = [a_{ij}]$ and $B = [b_{ij}]$ are $m \times n$ matrices, and let α denote a real number.

The *sum* of A and B is the matrix $C = A + B$ whose entries are

$$c_{ij} = a_{ij} + b_{ij}$$

for all i, j. The *difference* of A and B is the matrix $D = A - B$, where

$$d_{ij} = a_{ij} - b_{ij}$$

for all i, j. The *scalar multiple* of A (or the *product of a scalar* α *and a matrix* A) is the matrix $E = \alpha A$, where

$$e_{ij} = \alpha a_{ij}$$

for all i, j.

Only matrices of the same size can be added or subtracted. To multiply a matrix by a real number α, we multiply all of its entries by α. The product $(-1)A$ of the scalar -1 and a matrix A is denoted by $-A$. With this in mind, we can define subtraction in terms of addition: $A - B = A + (-B)$.

Example 8.1 Matrix Operations

Given the 2×2 matrices

$$A = \begin{bmatrix} 1 & 2 \\ 3 & 4 \end{bmatrix}, \quad B = \begin{bmatrix} 4 & 0 \\ 0 & 5 \end{bmatrix}, \quad \text{and} \quad C = \begin{bmatrix} -2 & 4 \\ 0 & 0 \end{bmatrix}$$

(a) compute the matrix $M = 3A - B + 7C$

(b) find the matrix $N = B - 4I_2$

(a) By repeatedly using Definition 22, we get

$$\begin{aligned} M &= 3A - B + 7C \\ &= 3\begin{bmatrix} 1 & 2 \\ 3 & 4 \end{bmatrix} - \begin{bmatrix} 4 & 0 \\ 0 & 5 \end{bmatrix} + 7\begin{bmatrix} -2 & 4 \\ 0 & 0 \end{bmatrix} \\ &= \begin{bmatrix} 3 & 6 \\ 9 & 12 \end{bmatrix} - \begin{bmatrix} 4 & 0 \\ 0 & 5 \end{bmatrix} + \begin{bmatrix} -14 & 28 \\ 0 & 0 \end{bmatrix} \\ &= \begin{bmatrix} 3-4-14 & 6-0+28 \\ 9-0+0 & 12-5+0 \end{bmatrix} = \begin{bmatrix} -15 & 34 \\ 9 & 7 \end{bmatrix} \end{aligned}$$

(b) Likewise,

$$\begin{aligned} N &= B - 4I_2 \\ &= \begin{bmatrix} 4 & 0 \\ 0 & 5 \end{bmatrix} - 4\begin{bmatrix} 1 & 0 \\ 0 & 1 \end{bmatrix} \\ &= \begin{bmatrix} 4 & 0 \\ 0 & 5 \end{bmatrix} - \begin{bmatrix} 4 & 0 \\ 0 & 4 \end{bmatrix} = \begin{bmatrix} 0 & 0 \\ 0 & 1 \end{bmatrix} \end{aligned}$$

Since matrices are added (subtracted) and multiplied by scalars component-wise, it is not surprising that the properties of these matrix operations are analogous to the properties of real numbers.

Theorem 6 **Properties of Matrix Addition and Multiplication by a Scalar**

Assume that A, B, and C are $m \times n$ matrices and let α and β denote real numbers. Then the following properties hold:

(a) *commutativity* of matrix addition: $A + B = B + A$

(b) *associativity* of matrix addition: $(A + B) + C = A + (B + C)$

(c) zero matrix: $A + 0 = A$

(d) cancellation with the negative of a matrix: $A + (-A) = 0$

(e) *distributivity* with respect to matrix addition: $\alpha(A + B) = \alpha A + \alpha B$

(f) *distributivity* with respect to scalar addition: $(\alpha + \beta)A = \alpha A + \beta A$

(g) *distributivity* with respect to multiplication of scalars: $\alpha(\beta A) = (\alpha\beta)A$

(h) scalar multiplication by 0 and 1: $0 \cdot A = 0$, $1 \cdot A = A$

The proofs of these statements consist of checking that the corresponding entries in the matrices on both sides of the equal sign are equal. For instance, the ijth entry on the left side of (e) is $\alpha(a_{ij} + b_{ij})$ and the corresponding entry on the right side is $\alpha a_{ij} + \alpha b_{ij}$. By the properties of real numbers, $\alpha(a_{ij} + b_{ij}) = \alpha a_{ij} + \alpha b_{ij}$.

In summary, the best way to remember Theorem 6 is to remember that matrix addition (subtraction) and multiplication of a matrix by a scalar share the properties of the real numbers.

Example 8.2 Solving a Matrix Equation

Define the matrices

$$A = \begin{bmatrix} 3 & 0 \\ 0 & 6 \\ 18 & 0 \end{bmatrix} \quad \text{and} \quad B = \begin{bmatrix} 1 & 0 \\ -4 & 1 \\ 9 & 2 \end{bmatrix}$$

Find the matrix X such that $5A - X = 2(3B + X)$.

▶ Using the properties listed in Theorem 6,

$$2(3B + X) = 2(3B) + 2X = 6B + 2X$$

and therefore

$$5A - X = 6B + 2X$$

Adding $-2X$ to both sides,

$$\begin{aligned} 5A - 3X &= 6B \\ -3X &= 6B - 5A \\ X &= -\frac{1}{3}(6B - 5A) \\ X &= -2B + \frac{5}{3}A \end{aligned}$$

Note that this is exactly how we would solve an equation in real numbers. (We skipped a number of steps in this calculation; if we wish to identify exactly which properties of Theorem 6 were used, we'd probably need twice as many steps.) It follows that

$$X = -2\begin{bmatrix} 1 & 0 \\ -4 & 1 \\ 9 & 2 \end{bmatrix} + \frac{5}{3}\begin{bmatrix} 3 & 0 \\ 0 & 6 \\ 18 & 0 \end{bmatrix}$$

$$X = \begin{bmatrix} -2 & 0 \\ 8 & -2 \\ -18 & -4 \end{bmatrix} + \begin{bmatrix} 5 & 0 \\ 0 & 10 \\ 30 & 0 \end{bmatrix} = \begin{bmatrix} 3 & 0 \\ 8 & 8 \\ 12 & -4 \end{bmatrix}$$

Definition 23 Transpose of a Matrix

Assume that $A = [a_{ij}]$ is an $m \times n$ matrix. The *transpose* A^t of A is the $n \times m$ matrix obtained by interchanging rows and columns of A.

In terms of matrix entries, we write $a^t_{ij} = a_{ji}$.

The transpose of A is also denoted by A'. If

$$A = \begin{bmatrix} 1 & 2 & 3 \\ 4 & 5 & 6 \\ 7 & 8 & 9 \end{bmatrix}, \quad B = \begin{bmatrix} 1 \\ 2 \\ 3 \end{bmatrix}, \quad C = [1 \quad 2], \quad \text{and} \quad D = \begin{bmatrix} 1 & 2 & 3 \\ 2 & 4 & 5 \\ 3 & 5 & 6 \end{bmatrix}$$

then

$$A^t = \begin{bmatrix} 1 & 4 & 7 \\ 2 & 5 & 8 \\ 3 & 6 & 9 \end{bmatrix}, \quad B^t = [1 \quad 2 \quad 3] \quad C^t = \begin{bmatrix} 1 \\ 2 \end{bmatrix} \quad \text{and} \quad D^t = \begin{bmatrix} 1 & 2 & 3 \\ 2 & 4 & 5 \\ 3 & 5 & 6 \end{bmatrix}$$

Note that $D^t = D$. A matrix that is equal to its transpose is called a *symmetric matrix*. For instance, the identity matrix is symmetric, and so is any square matrix whose off-diagonal terms are zero.

Transposition is a technical operation, that turns out to be quite helpful in various situations.

Matrix Multiplication

We now learn how to multiply matrices. The reasons why the product of matrices is defined in a way that's not at all straightforward will become clear as soon as we start applying matrices. For example, using matrix multiplication we will define an important class of functions. The composition of these functions corresponds to the product of matrices.

Definition 24 Matrix Multiplication

Assume that $A = [a_{ij}]$ is an $m \times n$ matrix and $B = [b_{ij}]$ is an $n \times p$ matrix. The *product* $C = AB$ of A and B is the $m \times p$ matrix $C = [c_{ij}]$ whose entries are given by

$$c_{ij} = a_{i1}b_{1j} + a_{i2}b_{2j} + \cdots + a_{in}b_{nj} \tag{8.2}$$

for $i = 1, 2, \ldots, m$ and $j = 1, 2, \ldots, p$.

The definition states that we cannot multiply matrices of arbitrary sizes: the product AB is defined only if the number of columns of the first factor (A) is equal to the number of rows of the second factor (B). Then the product AB is the matrix with the same number of rows as the first factor and the same number of columns as the second factor.

This probably sounds a bit confusing, so we restate it: if

$$A \text{ is an } m \times n \text{ matrix and } B \text{ is an } n \times p \text{ matrix,}$$

then the product AB is defined. The product of

$$\text{an } m \times n \text{ matrix } A \text{ and an } n \times p \text{ matrix } B$$

is an $m \times p$ matrix.

How do we calculate the entries in the product matrix? Looking at the right side of formula (8.2), we see that the first factors, $a_{i1}, a_{i2}, \ldots, a_{in}$, belong to the ith row of A:

$$[a_{i1} \quad a_{i2} \quad \cdots \quad a_{in}] \tag{8.3}$$

The second factors, $b_{1j}, b_{2j}, \ldots, b_{nj}$, are the entries in the jth column of B:

$$\begin{bmatrix} b_{1j} \\ b_{2j} \\ \vdots \\ b_{nj} \end{bmatrix} \tag{8.4}$$

Thinking of both (8.3) and (8.4) as vectors, formula (8.2) tells us that c_{ij} is the *dot product of the ith row of A and the jth column of B.*

Example 8.3 Multiplying Matrices

Let

$$A = \begin{bmatrix} 1 & 2 & -3 \\ 0 & 4 & 1 \end{bmatrix} \quad \text{and} \quad B = \begin{bmatrix} -2 & 1 \\ 5 & 2 \\ 0 & 6 \end{bmatrix}$$

Compute AB and BA, if defined.

Since A is a 2×3 matrix and B is a 3×2 matrix, the product matrix AB is defined; it is a (number of rows of A) $\times$ (number of columns of B), i.e., a 2×2 matrix.

To calculate the entry c_{11} of C, we compute the dot product of the first row of A and the first column of B:

$$c_{11} = \begin{bmatrix} 1 & 2 & -3 \end{bmatrix} \cdot \begin{bmatrix} -2 \\ 5 \\ 0 \end{bmatrix} = (1)(-2) + (2)(5) + (-3)(0) = -2 + 10 = 8$$

The entry c_{12} is the dot product of the first row of A and the second column of B,

$$c_{12} = \begin{bmatrix} 1 & 2 & -3 \end{bmatrix} \cdot \begin{bmatrix} 1 \\ 2 \\ 6 \end{bmatrix} = (1)(1) + (2)(2) + (-3)(6) = 1 + 4 - 18 = -13$$

In the same way we compute the remaining two entries:

$$c_{21} = \begin{bmatrix} 0 & 4 & 1 \end{bmatrix} \cdot \begin{bmatrix} -2 \\ 5 \\ 0 \end{bmatrix} = (0)(-2) + (4)(5) + (1)(0) = 20$$

$$c_{22} = \begin{bmatrix} 0 & 4 & 1 \end{bmatrix} \cdot \begin{bmatrix} 1 \\ 2 \\ 6 \end{bmatrix} = (0)(1) + (4)(2) + (1)(6) = 14$$

Thus,

$$C = AB = \begin{bmatrix} 8 & -13 \\ 20 & 14 \end{bmatrix}$$

The product $D = BA$ is defined as well, since B is a 3×2 matrix and A is a 2×3 matrix; D is a 3×3 matrix. Even without computing BA, we know that $AB \neq BA$, because the matrices are of different sizes.

To find $D = [d_{ij}]$, we need to calculate nine entries. For example, d_{23} is the dot product of the second row of B and the third column of A,

$$d_{23} = \begin{bmatrix} 5 & 2 \end{bmatrix} \cdot \begin{bmatrix} -3 \\ 1 \end{bmatrix} = -13$$

Likewise,

$$d_{31} = \begin{bmatrix} 0 & 6 \end{bmatrix} \cdot \begin{bmatrix} 1 \\ 0 \end{bmatrix} = 0$$

$$d_{22} = \begin{bmatrix} 5 & 2 \end{bmatrix} \cdot \begin{bmatrix} 2 \\ 4 \end{bmatrix} = 18$$

$$d_{13} = \begin{bmatrix} -2 & 1 \end{bmatrix} \cdot \begin{bmatrix} -3 \\ 1 \end{bmatrix} = 7$$

Calculating the remaining entries in the same way, we obtain

$$D = BA = \begin{bmatrix} -2 & 0 & 7 \\ 5 & 18 & -13 \\ 0 & 24 & 6 \end{bmatrix}$$

This example illustrates the fact that matrix multiplication is *not commutative.* It is not just that $AB \neq BA$, but it could happen that one of the products is not even defined. For instance, if A is a 3×4 matrix and B is a 4×2 matrix, then AB is a 4×2 matrix; however, the product BA is not defined.

Example 8.4 More Matrix Multiplication

Let

$$A = \begin{bmatrix} 1 & -3 \\ 2 & -6 \end{bmatrix} \quad \text{and} \quad B = \begin{bmatrix} 3 & -3 \\ 1 & -1 \end{bmatrix}$$

Compute AB and BA.

▶ Since A and B are 2×2 matrices, both AB and BA are defined and are of size 2×2. The entries in the matrix $C = AB$ are

$$c_{11} = \begin{bmatrix} 1 & -3 \end{bmatrix} \cdot \begin{bmatrix} 3 \\ 1 \end{bmatrix} = 0$$

$$c_{12} = \begin{bmatrix} 1 & -3 \end{bmatrix} \cdot \begin{bmatrix} -3 \\ -1 \end{bmatrix} = 0$$

$$c_{21} = \begin{bmatrix} 2 & -6 \end{bmatrix} \cdot \begin{bmatrix} 3 \\ 1 \end{bmatrix} = 0$$

$$c_{22} = \begin{bmatrix} 2 & -6 \end{bmatrix} \cdot \begin{bmatrix} -3 \\ -1 \end{bmatrix} = 0$$

Thus, C is the zero matrix. In the same way, we calculate the entries in $D = BA$:

$$d_{11} = \begin{bmatrix} 3 & -3 \end{bmatrix} \cdot \begin{bmatrix} 1 \\ 2 \end{bmatrix} = -3$$

$$d_{12} = \begin{bmatrix} 3 & -3 \end{bmatrix} \cdot \begin{bmatrix} -3 \\ -6 \end{bmatrix} = 9$$

$$d_{21} = \begin{bmatrix} 1 & -1 \end{bmatrix} \cdot \begin{bmatrix} 1 \\ 2 \end{bmatrix} = -1$$

$$d_{22} = \begin{bmatrix} 1 & -1 \end{bmatrix} \cdot \begin{bmatrix} -3 \\ -6 \end{bmatrix} = 3$$

Thus,

$$AB = \begin{bmatrix} 0 & 0 \\ 0 & 0 \end{bmatrix} \quad \text{and} \quad BA = \begin{bmatrix} -3 & 9 \\ -1 & 3 \end{bmatrix}$$

This is another example where $AB \neq BA$. Note that neither A nor B is a zero matrix, but their product AB is zero (another big difference between matrices and real numbers).

As usual, using multiplication, we define the *powers* of a square matrix A:

$$A^0 = I$$
$$A^1 = A$$
$$A^2 = A \cdot A$$
$$A^3 = A^2 \cdot A$$

and so on (I is the identity matrix.)

In Exercise 22 we show that the powers of the non-zero matrix

$$A = \begin{bmatrix} 0 & 1 & 2 \\ 0 & 0 & 3 \\ 0 & 0 & 0 \end{bmatrix}$$

satisfy $A^2 \neq 0$ and $A^3 = 0$.

Next, we list the properties of matrix multiplication. As we will see, some are analogous to the properties of real numbers.

Theorem 7 Properties of Matrix Multiplication

Assume that the matrices A, B, and C are of the appropriate sizes so that all operations in the formulas below are defined. Let α denote a real number. The following properties hold for matrix multiplication:

(a) *associativity:* $(AB)C = A(BC)$

(b) *distributivity* with respect to matrix addition: $(A + B)C = AC + BC$

(c) *distributivity* with respect to matrix addition: $A(B + C) = AB + AC$

(d) *distributivity* with respect to scalar multiplication: $(\alpha A)B = A(\alpha B) = \alpha(AB)$

(e) multiplication by the identity matrix: $AI = IA = A$

(f) multiplication by the zero matrix: $A0 = 0A = 0$

Since matrix multiplication is not commutative, neither of (b) or (c) follows from the other.

The proofs consist of straightforward calculations but can be quite messy (see Exercises 30 to 32 for the proofs for matrices of small sizes).

The Determinant of a Matrix

To any square matrix we assign a real number called the *determinant.* Although the determinant can be defined for a square matrix of any size, we define it only for 2×2 and 3×3 matrices.

Definition 25 The Determinant of a Matrix

The *determinant* of a 2×2 matrix

$$A = \begin{bmatrix} a_{11} & a_{12} \\ a_{21} & a_{22} \end{bmatrix}$$

is the real number

$$\det(A) = \begin{vmatrix} a_{11} & a_{12} \\ a_{21} & a_{22} \end{vmatrix} = a_{11}a_{22} - a_{12}a_{21} \tag{8.5}$$

The *determinant* of a 3×3 matrix

$$A = \begin{bmatrix} a_{11} & a_{12} & a_{13} \\ a_{21} & a_{22} & a_{23} \\ a_{31} & a_{32} & a_{33} \end{bmatrix}$$

is defined as

$$\det(A) = \begin{vmatrix} a_{11} & a_{12} & a_{13} \\ a_{21} & a_{22} & a_{23} \\ a_{31} & a_{32} & a_{33} \end{vmatrix}$$

$$= a_{11} \begin{vmatrix} a_{22} & a_{23} \\ a_{32} & a_{33} \end{vmatrix} - a_{12} \begin{vmatrix} a_{21} & a_{23} \\ a_{31} & a_{33} \end{vmatrix} + a_{13} \begin{vmatrix} a_{21} & a_{22} \\ a_{31} & a_{32} \end{vmatrix} \tag{8.6}$$

where the 2×2 determinants are computed as in (8.5).

Sometimes we drop the parentheses and write $\det A$ instead of $\det(A)$.

The determinant of a square matrix of order 2 is easy to compute: we multiply the diagonal entries and subtract the product of the off-diagonal entries:

$$\begin{vmatrix} 1 & 2 \\ 3 & 4 \end{vmatrix} = (1)(4) - (2)(3) = -2$$

or

$$\begin{vmatrix} 3 & 5 \\ -3 & -5 \end{vmatrix} = (3)(-5) - (-3)(5) = 0$$

How do we think of the 3×3 determinant in (8.6)?

The terms in front of the 2×2 determinants are the entries in the first row of A (with a minus sign placed in front of a_{12}, the reasons for which we will not go into). When we cross out the row and the column where each of the three entries is located, we obtain the corresponding 2×2 determinant.

Example 8.5 Calculating a 3×3 Determinant

Compute the determinant of the matrix

$$A = \begin{bmatrix} 2 & 3 & -4 \\ -2 & 0 & 1 \\ 5 & -1 & 2 \end{bmatrix}$$

▶ We see that $a_{11} = 2$. Crossing out the first row and the first column in A, we obtain the matrix

$$\begin{bmatrix} 0 & 1 \\ -1 & 2 \end{bmatrix}$$

whose determinant is

$$\begin{vmatrix} 0 & 1 \\ -1 & 2 \end{vmatrix} = (0)(2) - (-1)(1) = 1$$

Next, $a_{12} = 3$; the corresponding matrix is

$$\begin{bmatrix} -2 & 1 \\ 5 & 2 \end{bmatrix}$$

and its determinant is

$$\begin{vmatrix} -2 & 1 \\ 5 & 2 \end{vmatrix} = (-2)(2) - (5)(1) = -9$$

The determinant corresponding to $a_{13} = -4$ is

$$\begin{vmatrix} -2 & 0 \\ 5 & -1 \end{vmatrix} = (-2)(-1) - (5)(0) = 2$$

It follows that

$$A = 2\begin{vmatrix} 0 & 1 \\ -1 & 2 \end{vmatrix} - 3\begin{vmatrix} -2 & 1 \\ 5 & 2 \end{vmatrix} + (-4)\begin{vmatrix} -2 & 0 \\ 5 & -1 \end{vmatrix}$$

$$= 2(1) - 3(-9) + (-4)(2) = 21$$

We explore properties of determinants in Exercises 37 to 39.

Summary Matrices of the same size can be **added** and **subtracted.** As well, we can **multiply a matrix by a real number.** The properties of these operations mirror the properties of the real numbers. Two matrices can be **multiplied** if the number of columns of the first factor is equal to the number of rows of the second factor. Matrix multiplication is not commutative; as well, the product of two nonzero matrices can be zero. To **transpose a matrix,** we interchange its rows and columns. The **determinant** assigns a real number to a square matrix.

8 Exercises

1. What is the size of the matrix AB if A is a 1×3 matrix and B is a 3×1 matrix?

2. Let A be an $m \times n$ matrix. Is it true that the products AA^t and A^tA are always defined (i.e., for any value of m and n)?

3. Find an example of a non-zero 2×2 matrix whose square is a zero matrix.

4. Find an example of non-zero 2×2 matrices A and B whose product AB is a zero matrix.

5. Assume that A is a 3×3 matrix. Show that $A + A^t$ is a symmetric matrix.

6. Assume that A and B are 2×2 matrices. Is it true that $(A+B)^2 = A^2 + 2AB + B^2$?

7–20 ▪ Define the matrices

$$A = \begin{bmatrix} 1 & 2 \\ 0 & 4 \end{bmatrix}, \quad B = \begin{bmatrix} 4 & 0 \\ 0 & -2 \end{bmatrix}, \quad C = \begin{bmatrix} 3 & 1 & 0 \\ 1 & -2 & -1 \\ 4 & 0 & 3 \end{bmatrix}, \quad D = \begin{bmatrix} 0 & 3 & 2 \\ 0 & 0 & 6 \\ 0 & 0 & 0 \end{bmatrix}$$

$$E = \begin{bmatrix} 1 & 2 \\ 3 & 0 \\ 2 & 5 \end{bmatrix}, \quad F = \begin{bmatrix} 1 & 0 & 0 \\ 1 & 2 & 3 \end{bmatrix}, \quad G = \begin{bmatrix} 5 \\ 6 \end{bmatrix}, \quad H = \begin{bmatrix} 1 \\ -2 \\ 3 \end{bmatrix}$$

In each case, determine whether or not the matrix operations are defined. If so, find the matrix.

7. $7A - 2B + I_2$
8. $3C - 5E$
9. $3E - 4F^t$
10. $I_3 - 0.5D^t$
11. AE
12. EA
13. $D - E^2$
14. CC^t
15. $D - EF$
16. $CH - H$
17. D^3
18. $A^2 + BB^t$
19. BGG^t
20. $HG^t + E$

21. A matrix of the form

$$S = \begin{bmatrix} s & 0 \\ 0 & s \end{bmatrix}$$

where s is a real number, is called a *scalar matrix*. Show that $AS = SA$ for any 2×2 matrix A. Formulate a rule that states how a matrix can be multiplied by a scalar matrix.

22. Let

$$A = \begin{bmatrix} 0 & 1 & 2 \\ 0 & 0 & 3 \\ 0 & 0 & 0 \end{bmatrix}$$

Show that $A^2 = A \cdot A$ has a single non-zero entry. Verify that $A^3 = A^2 \cdot A = 0$.

23. A matrix of the form

$$D = \begin{bmatrix} d_1 & 0 & 0 \\ 0 & d_2 & 0 \\ 0 & 0 & d_3 \end{bmatrix}$$

where d_1, d_2, and d_3 are real numbers, is called a *diagonal matrix*. Let A be a 3×3 matrix. Explain how the entries of AD and DA are obtained.

24. A matrix of the form

$$S = \begin{bmatrix} s & 0 & 0 \\ 0 & s & 0 \\ 0 & 0 & s \end{bmatrix}$$

where s is a real number, is called a *scalar matrix*. Let A be a 3×3 matrix. Explain how the entries of AS and SA are obtained.

25. Assume that θ is a real number, and define

$$A = \begin{bmatrix} \cos\theta & -\sin\theta \\ \sin\theta & \cos\theta \end{bmatrix}$$

Show that

$$A^2 = \begin{bmatrix} \cos 2\theta & -\sin 2\theta \\ \sin 2\theta & \cos 2\theta \end{bmatrix}$$

26–29 ▪ Define the matrices

$$A = \begin{bmatrix} -1 & 2 \\ 0 & 4 \end{bmatrix}, \quad B = \begin{bmatrix} -2 & 0 \\ 0 & -2 \end{bmatrix}, \quad C = \begin{bmatrix} 8 & 1 \\ 5 & 1 \end{bmatrix}, \quad D = \begin{bmatrix} 0 & 0 \\ 2 & 0 \end{bmatrix}$$

Solve each equation for X.

26. $2X - (A + 2B) = I_2 - X$

27. $4(X + D) = BC$

28. $C - 3(I_2 - X) = 4(X + D)$

29. $5X - A = X - 5C$

30. Show that $(A + B)C = AC + BC$ holds for 2×2 matrices.

31. Show that $(\alpha A)B = \alpha(AB)$ holds for 2×2 matrices.

32. Assume that A is a 3×3 matrix; I is the identity matrix of size 3. Show that $AI = IA = A$.

33–36 ▪ Find the determinant of each matrix.

33. $\begin{bmatrix} 1 & -2 \\ -3 & 4 \end{bmatrix}$

34. $\begin{bmatrix} 0 & 10 \\ 1 & 10 \end{bmatrix}$

35. $\begin{bmatrix} 1 & 2 & 3 \\ 2 & 3 & 4 \\ 3 & 5 & 7 \end{bmatrix}$

36. $\begin{bmatrix} 2 & 1 & 3 \\ 0 & 2 & 0 \\ -3 & 0 & 6 \end{bmatrix}$

37. Assume that A is a 2×2 matrix and c is a real number. How are the determinants of A and cA related?

38. Assume that A is a 3×3 matrix and c is a real number. How are the determinants of A and cA related?

39. Let

$$A = \begin{bmatrix} -1 & 3 \\ 1 & 4 \end{bmatrix} \quad \text{and} \quad B = \begin{bmatrix} 2 & 3 \\ 1 & 1 \end{bmatrix}$$

Convince yourself that $AB \neq BA$, but $\det(AB) = \det(BA)$.

9 Matrices and Linear Systems

Using **inverse matrices,** we develop an alternative way of solving certain systems of linear equations.

Introduction

In Section 6, to a system of linear equations

$$\begin{aligned} a_{11}x_1 + a_{12}x_2 + \cdots + a_{1n}x_n &= b_1 \\ a_{21}x_1 + a_{22}x_2 + \cdots + a_{2n}x_n &= b_2 \\ &\cdots \\ a_{m1}x_1 + a_{m2}x_2 + \cdots + a_{mn}x_n &= b_m \end{aligned}$$

we associated the coefficient matrix (matrix of the system)

$$A = \begin{bmatrix} a_{11} & a_{12} & \cdots & a_{1n} \\ a_{21} & a_{22} & \cdots & a_{2n} \\ \cdots & \cdots & \cdots & \cdots \\ a_{m1} & a_{m2} & \cdots & a_{mn} \end{bmatrix} \tag{9.1}$$

and the matrix of constant terms (free coefficients)

$$\mathbf{b} = \begin{bmatrix} b_1 \\ b_2 \\ \cdots \\ b_m \end{bmatrix}$$

Defining the matrix of variables

$$\mathbf{x} = \begin{bmatrix} x_1 \\ x_2 \\ \cdots \\ x_n \end{bmatrix}$$

allows us to write (9.1) in matrix form:

$$A\mathbf{x} = \mathbf{b}$$

Since A is an $m \times n$ matrix and $\mathbf{x}$ is an $n \times 1$ matrix, the product matrix on the left side is an $m \times 1$ matrix, of the same type as the matrix $\mathbf{b}$ on the right side.

In this section, we assume that $m = n$, i.e., we assume that the system has the same number of equations and variables (so it's neither over-determined nor under-determined). This assumption implies that A is a square matrix.

Example 9.1 · Linear System in Matrix Form

The system

$$\begin{aligned} x - 2y + 3z &= 5 \\ 3x + y &= -5 \\ -2x + y + z &= 8 \end{aligned} \tag{9.2}$$

of Example 6.1 in Section 6 can be written as $A\mathbf{x} = \mathbf{b}$, where

$$A = \begin{bmatrix} 1 & -2 & 3 \\ 3 & 1 & 0 \\ -2 & 1 & 1 \end{bmatrix}, \quad \mathbf{x} = \begin{bmatrix} x \\ y \\ z \end{bmatrix}, \quad \text{and} \quad \mathbf{b} = \begin{bmatrix} 5 \\ -5 \\ 8 \end{bmatrix}$$

The equation

$$A\mathbf{x} = \mathbf{b} \tag{9.3}$$

is an example of a *matrix equation;* the unknown variable is the $n \times 1$ matrix $\mathbf{x}$. Equation (9.3) resembles the equation

$$ax = b \tag{9.4}$$

in real numbers. To solve (9.4) for x, we multiply both sides by $a^{-1} = 1/a$ (assuming that $a \neq 0$; if $a = 0$, we don't have an equation):

$$\begin{aligned} a^{-1} \cdot ax &= a^{-1} \cdot b \\ 1 \cdot x &= \frac{1}{a} b \\ x &= \frac{b}{a} \end{aligned}$$

So, to find x from $ax = b$, we divide by a. Going back to matrices, given that $A\mathbf{x} = \mathbf{b}$, how do we divide by the matrix A, i.e., how do we make sense of

$$\mathbf{x} = \frac{\mathbf{b}}{A}$$

It turns out that under certain conditions (which are equivalent to the requirement $a \neq 0$ in (9.4)), a *square* matrix A has an *inverse:* i.e., there is a matrix A^{-1} such that

$$A \cdot A^{-1} = A^{-1} \cdot A = I$$

where I denotes the identity matrix. In order to solve the matrix equation $A\mathbf{x} = \mathbf{b}$, we multiply both sides from the left by A^{-1} (keep in mind that matrix multiplication is not commutative):

$$\begin{aligned} A^{-1} \cdot A\mathbf{x} &= A^{-1} \cdot \mathbf{b} \\ I \cdot \mathbf{x} &= A^{-1}\mathbf{b} \\ \mathbf{x} &= A^{-1}\mathbf{b} \end{aligned}$$

In other words, if we can find an inverse matrix for the coefficient matrix A, then the solution of the system $A\mathbf{x} = \mathbf{b}$ is given by the matrix multiplication $\mathbf{x} = A^{-1}\mathbf{b}$. (Thus, the "division" $\mathbf{b}/A$ amounts to the product of A^{-1} and $\mathbf{b}$.)

Example 9.2 **Example 9.1, Continued**

Soon, we will be able to show that the inverse of the matrix A in Example 9.1 is

$$A^{-1} = \begin{bmatrix} 1/22 & 5/22 & -3/22 \\ -3/22 & 7/22 & 9/22 \\ 5/22 & 3/22 & 7/22 \end{bmatrix}$$

Thus, the solution of the system (9.2) is given by

$$\begin{aligned} \begin{bmatrix} x \\ y \\ z \end{bmatrix} &= A^{-1}\mathbf{b} \\ &= \begin{bmatrix} 1/22 & 5/22 & -3/22 \\ -3/22 & 7/22 & 9/22 \\ 5/22 & 3/22 & 7/22 \end{bmatrix} \begin{bmatrix} 5 \\ -5 \\ 8 \end{bmatrix} \\ &= \begin{bmatrix} (1/22)(5) + (5/22)(-5) - (3/22)(8) \\ (-3/22)(5) + (7/22)(-5) + (9/22)(8) \\ (5/22)(5) + (3/22)(-5) + (7/22)(8) \end{bmatrix} = \begin{bmatrix} -2 \\ 1 \\ 3 \end{bmatrix} \end{aligned}$$

i.e., $x = -2$, $y = 1$, and $z = 3$, confirming our calculations using Gaussian elimination in Example 6.1. ▲

The Inverse of a Matrix

We define inverse matrices, study their properties, and learn how to find them.

Definition 26 The Inverse of a Matrix

Assume that A is a square matrix of size n. An $n \times n$ matrix B satisfying

$$AB = BA = I$$

where I is the identity matrix, is called an *inverse matrix* of A.

Note the symmetry implied by the definition: if B is an inverse matrix of A, then A is an inverse matrix of B as well.

An inverse of A is usually denoted by A^{-1}. If A has an inverse matrix, then

$$AA^{-1} = A^{-1}A = I$$

Because matrix multiplication is not commutative, we have to require that the product of A and A^{-1} in either order be equal to the identity matrix.

A matrix that has an inverse matrix is called *invertible,* or *non-singular.* Otherwise, if a matrix does not have an inverse matrix, it is *non-invertible,* or *singular.*

Example 9.3 Inverse Matrix

Verify that the matrix

$$A = \begin{bmatrix} 3 & -2 \\ -1 & 4 \end{bmatrix}$$

is invertible by showing that

$$A^{-1} = \begin{bmatrix} 4/10 & 2/10 \\ 1/10 & 3/10 \end{bmatrix}$$

▶ Multiplying the two matrices, we obtain

$$\begin{aligned} AA^{-1} &= \begin{bmatrix} 3 & -2 \\ -1 & 4 \end{bmatrix} \begin{bmatrix} 4/10 & 2/10 \\ 1/10 & 3/10 \end{bmatrix} \\ &= \begin{bmatrix} 3(4/10) - 2(1/10) & 3(2/10) - 2(3/10) \\ -1(4/10) + 4(1/10) & -1(2/10) + 4(3/10) \end{bmatrix} \\ &= \begin{bmatrix} 1 & 0 \\ 0 & 1 \end{bmatrix} \end{aligned}$$

In the same way we show that $A^{-1}A = I$.

Example 9.4 Matrix Manipulation

Check that the matrix

$$A = \begin{bmatrix} 1 & 3 \\ 0 & -1 \end{bmatrix}$$

satisfies $A^2 = I$ (recall that $A^2 = A \cdot A$). What does this say about A^{-1}?

▶ We compute

$$A^2 = \begin{bmatrix} 1 & 3 \\ 0 & -1 \end{bmatrix} \begin{bmatrix} 1 & 3 \\ 0 & -1 \end{bmatrix} = \begin{bmatrix} 1 \cdot 1 + 3 \cdot 0 & 1 \cdot 3 + 3 \cdot (-1) \\ 0 \cdot 1 + (-1) \cdot 0 & 0 \cdot 3 + (-1) \cdot (-1) \end{bmatrix} = \begin{bmatrix} 1 & 0 \\ 0 & 1 \end{bmatrix}$$

So, indeed, $A \cdot A = I$; the matrix A multiplied by itself gives the identity matrix. Thus, $A^{-1} = A$; i.e., A is its own inverse matrix.

Example 9.5 Example 9.2, Revisited

Check that the matrix

$$A^{-1} = \begin{bmatrix} 1/22 & 5/22 & -3/22 \\ -3/22 & 7/22 & 9/22 \\ 5/22 & 3/22 & 7/22 \end{bmatrix}$$

is indeed an inverse of the matrix

$$A = \begin{bmatrix} 1 & -2 & 3 \\ 3 & 1 & 0 \\ -2 & 1 & 1 \end{bmatrix}$$

as claimed in Example 9.2.

▶ This is a straightforward exercise: we need to verify that $AA^{-1} = I$ and $A^{-1}A = I$. Below, we show the calculations in the first column; the remaining entries are found in the same way.

$$\begin{aligned} AA^{-1} &= \begin{bmatrix} 1 & -2 & 3 \\ 3 & 1 & 0 \\ -2 & 1 & 1 \end{bmatrix} \begin{bmatrix} 1/22 & 5/22 & -3/22 \\ -3/22 & 7/22 & 9/22 \\ 5/22 & 3/22 & 7/22 \end{bmatrix} \\ &= \begin{bmatrix} 1(1/22) - 2(-3/22) + 3(5/22) & 0 & 0 \\ 3(1/22) + 1(-3/22) + 0(5/22) & 1 & 0 \\ -2(1/22) + 1(-3/22) + 1(5/22) & 0 & 1 \end{bmatrix} \\ &= \begin{bmatrix} 1 & 0 & 0 \\ 0 & 1 & 0 \\ 0 & 0 & 1 \end{bmatrix} \end{aligned}$$

The remaining identity $A^{-1}A = I$ is checked analogously. ▲

We now discuss several properties of inverse matrices.

(a) If A^{-1} is an inverse matrix of A, then

$$AA^{-1} = A^{-1}A = I \tag{9.5}$$

Note that (9.5) implies that A is an inverse of A^{-1}, i.e.,

$$(A^{-1})^{-1} = A$$

(this is analogous to the fact that $1/(1/a) = a$ for non-zero real numbers).

(b) Since $I \cdot I = I$, the identity matrix is its own inverse; thus $I^{-1} = I$ (mirrors the fact that the real number 1 is its own reciprocal).

(c) Assume that A has two inverse matrices, B and C. Definition 26 implies that $AB = BA = I$ and $AC = CA = I$. Then

$$B = BI = B(AC) = (BA)C = IC = C$$

We conclude that an inverse matrix, if it exists, is unique (so we say "the inverse matrix of a matrix" instead of "an inverse matrix of a matrix").

(d) What is the inverse matrix of the product AB of two non-singular matrices? Consider the matrix $B^{-1}A^{-1}$:

$$(AB)(B^{-1}A^{-1}) = A(BB^{-1})A^{-1} = AIA^{-1} = AA^{-1} = I$$

Likewise, $(B^{-1}A^{-1})(AB) = I$ (see Exercise 26). We conclude that

$$(AB)^{-1} = B^{-1}A^{-1}$$

How do we find an inverse matrix?

Before introducing a general formula, we explore an example.

Example 9.6 Finding an Inverse Matrix

Find A^{-1} if

$$A = \begin{bmatrix} 1 & -3 \\ 1 & 4 \end{bmatrix}$$

▶ We are looking for the matrix

$$B = \begin{bmatrix} x & y \\ z & w \end{bmatrix}$$

such that $AB = BA = I$. (To simplify the notation, we use x, y, z, and w for the entries of B, instead of the lowercase b with double subscripts.)

Our strategy is to find B so that $AB = I$ and then check that $BA = I$ holds as well. From

$$AB = \begin{bmatrix} 1 & -3 \\ 1 & 4 \end{bmatrix} \begin{bmatrix} x & y \\ z & w \end{bmatrix} = \begin{bmatrix} 1 & 0 \\ 0 & 1 \end{bmatrix}$$

$$\begin{bmatrix} x - 3z & y - 3w \\ x + 4z & y + 4w \end{bmatrix} = \begin{bmatrix} 1 & 0 \\ 0 & 1 \end{bmatrix}$$

we obtain a pair of linear systems

$$\begin{aligned} x - 3z &= 1 \\ x + 4z &= 0 \end{aligned} \tag{9.6}$$

and

$$\begin{aligned} y - 3w &= 0 \\ y + 4w &= 1 \end{aligned} \tag{9.7}$$

The augmented matrix for the system (9.6) is

$$\left[\begin{array}{cc|c} 1 & -3 & 1 \\ 1 & 4 & 0 \end{array}\right] \begin{array}{l} (R_1) \\ (R_2) \end{array}$$

By using elementary row operations, we reduce it to row echelon form:

$$\left[\begin{array}{cc|c} 1 & -3 & 1 \\ 0 & 7 & -1 \end{array}\right] \begin{array}{l} (R_3 \leftarrow R_1) \\ (R_4 \leftarrow -R_1 + R_2) \end{array}$$

We could get the solutions from here by using back substitution. However, we continue, obtaining an even simpler equivalent form (the reason we are doing this will become clear later in this section)

$$\left[\begin{array}{cc|c} 1 & -3 & 1 \\ 0 & 1 & -1/7 \end{array}\right] \begin{array}{l} (R_5 \leftarrow R_3) \\ (R_6 \leftarrow R_4/7) \end{array}$$

$$\left[\begin{array}{cc|c} 1 & 0 & 4/7 \\ 0 & 1 & -1/7 \end{array}\right] \begin{array}{l} (R_7 \leftarrow 3R_6 + R_5) \\ (R_8 \leftarrow R_6) \end{array} \tag{9.8}$$

With this form (called the *reduced row echelon form*), we do not need back substitution; the solutions are $x = 4/7$ and $z = -1/7$. Note that the solutions form the right-most column in (9.8); we will need this fact later.

To solve the system (9.7), we simplify its augmented matrix in the same way:

$$\left[\begin{array}{cc|c} 1 & -3 & 0 \\ 1 & 4 & 1 \end{array}\right] \begin{array}{l} (R_1) \\ (R_2) \end{array}$$

$$\left[\begin{array}{cc|c} 1 & -3 & 0 \\ 0 & 7 & 1 \end{array}\right] \begin{array}{l} (R_3 \leftarrow R_1) \\ (R_4 \leftarrow -R_1 + R_2) \end{array}$$

$$\left[\begin{array}{cc|c} 1 & -3 & 0 \\ 0 & 1 & 1/7 \end{array}\right] \begin{array}{l} (R_5 \leftarrow R_3) \\ (R_6 \leftarrow R_4/7) \end{array}$$

$$\left[\begin{array}{cc|c} 1 & 0 & 3/7 \\ 0 & 1 & 1/7 \end{array}\right] \begin{array}{l} (R_7 \leftarrow 3R_6 + R_5) \\ (R_8 \leftarrow R_6) \end{array}$$

The solutions are $y = 3/7$ and $w = 1/7$. Thus,

$$B = \begin{bmatrix} 4/7 & 3/7 \\ -1/7 & 1/7 \end{bmatrix}$$

Since

$$BA = \begin{bmatrix} 4/7 & 3/7 \\ -1/7 & 1/7 \end{bmatrix} \begin{bmatrix} 1 & -3 \\ 1 & 4 \end{bmatrix} = \begin{bmatrix} 1 & 0 \\ 0 & 1 \end{bmatrix}$$

it follows that B is indeed the inverse of A. Thus

$$A^{-1} = \begin{bmatrix} 4/7 & 3/7 \\ -1/7 & 1/7 \end{bmatrix}$$

Let us look a bit closer at the calculation in Example 9.6. The augmented matrices for the two systems (9.6) and (9.7)

$$\left[\begin{array}{cc|c} 1 & -3 & 1 \\ 1 & 4 & 0 \end{array}\right] \quad \text{and} \quad \left[\begin{array}{cc|c} 1 & -3 & 0 \\ 1 & 4 & 1 \end{array}\right]$$

have the same matrix on the left. Thus, it is not surprising that we used identical row operations in both cases.

To remove redundancy, we create the augmented matrix

$$\left[\begin{array}{cc|cc} 1 & -3 & 1 & 0 \\ 1 & 4 & 0 & 1 \end{array}\right]$$

which allows us to solve *both* systems at the same time. Using elementary row operations (as in Example 9.6, but only once!), we arrive at the matrix

$$\left[\begin{array}{cc|cc} 1 & 0 & 4/7 & 3/7 \\ 0 & 1 & -1/7 & 1/7 \end{array}\right]$$

The matrix on the right is the inverse matrix.

Algorithm 2 How to Find an Inverse Matrix

Assume that A is a square matrix.

(1) Form the augmented matrix $[\,A\,|\,I\,]$, where A is the given matrix and I is the identity matrix of the same size as A.

(2) Using elementary row operations, reduce the augmented matrix (if possible) to the form $[\,I\,|\,B\,]$. The matrix B is the inverse matrix of A.

(3) If the augmented matrix $[\,A\,|\,I\,]$ cannot be reduced to the form $[\,I\,|\,B\,]$, then A does not have an inverse matrix.

Note that reducing the augmented matrix from step (1) to the form $[\,I\,|\,B\,]$ finds the matrix B such that $AB = I$. It can be shown that B also satisfies $BA = I$, i.e., that B is indeed the inverse matrix of A, as claimed in step (2). (We will show soon that this is true for a general 2×2 matrix.)

Example 9.7 Identifying a Non-invertible Matrix

Find, if it exists, the inverse matrix of

$$A = \begin{bmatrix} 2 & -1 \\ -4 & 2 \end{bmatrix}$$

using Algorithm 2.

▶ We form the augmented matrix

$$\left[\begin{array}{cc|cc} 2 & -1 & 1 & 0 \\ -4 & 2 & 0 & 1 \end{array}\right] \begin{array}{c} (R_1) \\ (R_2) \end{array}$$

and work on the left-most column:

$$\left[\begin{array}{cc|cc} 1 & -1/2 & 1/2 & 0 \\ 0 & 0 & 2 & 1 \end{array}\right] \begin{array}{l} (R_3 \leftarrow R_1/2) \\ (R_4 \leftarrow 2R_1 + R_2) \end{array}$$

No matter what row operation we do, we cannot introduce a 1 in the second row and the second column. (In the language of linear systems of equations, the two systems represented by this augmented matrix are inconsistent.)

We conclude that the given matrix does not have an inverse matrix. ▲

We now apply Algorithm 2 find the inverse matrix of a general 2×2 matrix

$$A = \begin{bmatrix} a & b \\ c & d \end{bmatrix}$$

We form the augmented matrix

$$\left[\begin{array}{cc|cc} a & b & 1 & 0 \\ c & d & 0 & 1 \end{array}\right] \begin{array}{l} (R_1) \\ (R_2) \end{array}$$

First, transform the entry in the first column and the second row to zero. Assuming that $a \neq 0$,

$$\left[\begin{array}{cc|cc} a & b & 1 & 0 \\ 0 & d - bc/a & -c/a & 1 \end{array}\right] \begin{array}{l} (R_3 \leftarrow R_1) \\ (R_4 \leftarrow -(c/a)R_1 + R_2) \end{array}$$

Next, simplify and transform the entry in the second column and the second row into 1 (assume that $ad - bc \neq 0$):

$$\left[\begin{array}{cc|cc} a & b & 1 & 0 \\ 0 & ad - bc & -c & a \end{array}\right] \begin{array}{l} (R_5 \leftarrow R_3) \\ (R_6 \leftarrow aR_4) \end{array}$$

$$\left[\begin{array}{cc|cc} a & b & 1 & 0 \\ 0 & 1 & -c/(ad - bc) & a/(ad - bc) \end{array}\right] \begin{array}{l} (R_7 \leftarrow R_5) \\ (R_8 \leftarrow R_6/(ad - bc)) \end{array}$$

Introduce a zero in the first row and the second column and simplify the fraction in the first row:

$$\left[\begin{array}{cc|cc} a & 0 & 1 + bc/(ad - bc) & -ab/(ad - bc) \\ 0 & 1 & -c/(ad - bc) & a/(ad - bc) \end{array}\right] \begin{array}{l} (R_9 \leftarrow -bR_8 + R_7) \\ (R_{10} \leftarrow R_8) \end{array}$$

$$\left[\begin{array}{cc|cc} a & 0 & ad/(ad - bc) & -ab/(ad - bc) \\ 0 & 1 & -c/(ad - bc) & a/(ad - bc) \end{array}\right] \begin{array}{l} (R_{11}) \\ (R_{12}) \end{array}$$

Finally, divide the first row by a:

$$\left[\begin{array}{cc|cc} 1 & 0 & d/(ad - bc) & -b/(ad - bc) \\ 0 & 1 & -c/(ad - bc) & a/(ad - bc) \end{array}\right] \begin{array}{l} (R_{13} \leftarrow R_{11}/a) \\ (R_{14} \leftarrow R_{12}) \end{array}$$

Thus, the matrix

$$B = \begin{bmatrix} d/(ad - bc) & -b/(ad - bc) \\ -c/(ad - bc) & a/(ad - bc) \end{bmatrix}$$

satisfies $AB = I$. A straightforward calculation (see Exercise 6) shows that $BA = I$, and therefore B is the inverse of A:

$$\begin{bmatrix} a & b \\ c & d \end{bmatrix}^{-1} = \begin{bmatrix} d/(ad - bc) & -b/(ad - bc) \\ -c/(ad - bc) & a/(ad - bc) \end{bmatrix} = \frac{1}{ad - bc} \begin{bmatrix} d & -b \\ -c & a \end{bmatrix}$$

The only assumption that we need for this to work is that $ad - bc \neq 0$.

Note that $ad - bc$ is the determinant of A. Thus, if the determinant of A is not zero, then A is invertible.

Theorem 8 The Inverse of a 2×2 Matrix

Let

$$A = \begin{bmatrix} a & b \\ c & d \end{bmatrix}$$

and assume that $\det A = ad - bc \neq 0$. Then A is invertible, and

$$A^{-1} = \frac{1}{\det A} \begin{bmatrix} d & -b \\ -c & a \end{bmatrix}$$

Comparing the entries of A^{-1} with the entries of A, we realize that A^{-1} is easy to find: we switch the diagonal entries of A, change the signs of the off-diagonal entries, and divide all terms by the determinant of A.

For example,

$$\begin{bmatrix} 10 & -2 \\ 3 & 1 \end{bmatrix}^{-1} = \frac{1}{16} \begin{bmatrix} 1 & 2 \\ -3 & 10 \end{bmatrix}$$

(since $\det A = (10)(1) - (-2)(3) = 16$).

Unfortunately, the inverses of larger matrices are more difficult to find. As an illustration, we calculate the inverse of a 3×3 matrix. We will not concern ourselves with the inverses of matrices of size larger than 3.

Example 9.8 Finding the Inverse of a 3×3 Matrix

Find A^{-1} if

$$A = \begin{bmatrix} 1 & -2 & 3 \\ 3 & 1 & 0 \\ -2 & 1 & 1 \end{bmatrix}$$

(thus showing how the inverse matrix in Example 9.2 was obtained).

▶ We use Algorithm 2: form the augmented matrix

$$\left[\begin{array}{ccc|ccc} 1 & -2 & 3 & 1 & 0 & 0 \\ 3 & 1 & 0 & 0 & 1 & 0 \\ -2 & 1 & 1 & 0 & 0 & 1 \end{array}\right] \begin{array}{l} (R_1) \\ (R_2) \\ (R_3) \end{array}$$

and use elementary row operations to obtain the 3×3 identity matrix on the right. We work on the leftmost column first:

$$\left[\begin{array}{ccc|ccc} 1 & -2 & 3 & 1 & 0 & 0 \\ 0 & 7 & -9 & -3 & 1 & 0 \\ 0 & -3 & 7 & 2 & 0 & 1 \end{array}\right] \begin{array}{l} (R_4 \leftarrow R_1) \\ (R_5 \leftarrow -3R_1 + R_2) \\ (R_6 \leftarrow 2R_1 + R_3) \end{array}$$

Next, we use the leading entry 7 in the second row to eliminate the entries above and below it:

$$\left[\begin{array}{ccc|ccc} 1 & 0 & 3/7 & 1/7 & 2/7 & 0 \\ 0 & 7 & -9 & -3 & 1 & 0 \\ 0 & 0 & 22/7 & 5/7 & 3/7 & 1 \end{array}\right] \begin{array}{l} (R_7 \leftarrow (2/7)R_5 + R_4) \\ (R_8 \leftarrow R_5) \\ (R_9 \leftarrow (3/7)R_5 + R_6) \end{array}$$

Make the leading entries in the second and the third rows equal to 1:

$$\left[\begin{array}{ccc|ccc} 1 & 0 & 3/7 & 1/7 & 2/7 & 0 \\ 0 & 1 & -9/7 & -3/7 & 1/7 & 0 \\ 0 & 0 & 1 & 5/22 & 3/22 & 7/22 \end{array}\right] \begin{array}{l} (R_{10} \leftarrow R_7) \\ (R_{11} \leftarrow (1/7)R_8) \\ (R_{12} \leftarrow (7/22)R_9) \end{array}$$

Finally, we use the leading 1 in the third row to remove the entries above it:

$$\left[\begin{array}{ccc|ccc} 1 & 0 & 0 & 1/22 & 5/22 & -3/22 \\ 0 & 1 & 0 & -3/22 & 7/22 & 9/22 \\ 0 & 0 & 1 & 5/22 & 3/22 & 7/22 \end{array}\right] \begin{array}{l} (R_{13} \leftarrow -(3/7)R_{12} + R_{10}) \\ (R_{14} \leftarrow (9/7)R_{12} + R_{11}) \\ (R_{15} \leftarrow R_{12}) \end{array}$$

We are done:

$$A^{-1} = \begin{bmatrix} 1/22 & 5/22 & -3/22 \\ -3/22 & 7/22 & 9/22 \\ 5/22 & 3/22 & 7/22 \end{bmatrix}$$

This calculation is quite involved. Luckily, mathematical software and some calculators have built-in routines for finding inverse matrices.

Using Matrices to Solve Linear Systems

In conclusion, we know two ways to find the solutions of a linear system of equations:

(1) Form the augmented matrix of the system and, with the help of elementary row operations, reduce to row echelon form. Using back substitution, find the values of all variables.

(2) Write the system in matrix form as $A\mathbf{x} = \mathbf{b}$ and find the inverse matrix A^{-1}. The solution is given by $\mathbf{x} = A^{-1}\mathbf{b}$.

Note that (1) applies to all systems, no matter what the size or the number of solutions. Method (2) applies *only* to $n \times n$ systems (i.e., the same number of variables and equations), and *only* when the inverse matrix A^{-1} exists (in which case the solution is unique).

A generalization of Theorem 8 is given in the following statement.

Theorem 9 Non-singularity and the Determinant of a Matrix

Let A be an $n \times n$ matrix, where $n \geq 2$. Then A is non-singular (invertible) if and only if the determinant of A is not zero.

We know how to calculate the determinant of a 3×3 matrix. Because we will not need it, we do not define the determinant in general, for $n \times n$ matrices. Unfortunately, there are no easy formulas (analogous to the one in Theorem 8) for inverses of matrices of size larger than 2×2.

A special case of a general system of linear equations $A\mathbf{x} = \mathbf{b}$ is the *homogeneous* $n \times n$ linear system

$$A\mathbf{x} = \mathbf{0} \tag{9.9}$$

where $\mathbf{0}$ denotes a zero vector. ("Homogeneous" refers to the fact that all constant terms (free coefficients) are zero.)

Note that $\mathbf{x} = \mathbf{0}$ is a solution of (9.9), since $A\mathbf{0} = \mathbf{0}$ no matter what A is. This solution is called a *trivial solution.* If A is invertible, then $\mathbf{x} = A^{-1}\mathbf{0} = \mathbf{0}$ is the only solution of the system (9.9).

A solution $\mathbf{x}$ of (9.9) is called a *non-trivial solution* if $\mathbf{x} \neq \mathbf{0}$. The only way to obtain a non-trivial solution from a homogeneous system $A\mathbf{x} = \mathbf{0}$ is to use a singular matrix A. In conclusion, we have the following result.

Theorem 10 Solutions of a Homogeneous Linear System

Let A be an $n \times n$ matrix, $n \geq 2$. The homogeneous system of linear equations

$$A\mathbf{x} = \mathbf{0}$$

has a non-trivial solution if and only if A is singular (i.e., non-invertible).

Keep in mind that a matrix A is singular if and only if its determinant is zero.

Example 9.9 Homogeneous Linear Systems

In each case, solve the system $A\mathbf{x} = \mathbf{0}$:

(a) $A = \begin{bmatrix} 1 & 2 \\ -1 & 1 \end{bmatrix}$ (b) $A = \begin{bmatrix} 1 & 2 \\ -2 & -4 \end{bmatrix}$

▶ (a) The corresponding system is

$$x + 2y = 0$$
$$-x + y = 0$$

Adding the two equations, we get $3y = 0$ and $y = 0$. Either of the two equations implies that $x = 0$. Thus, the system has only a trivial solution.

Alternatively, $\det A = (1)(1) - (-1)(2) = 3 \neq 0$. Theorem 10 implies that the trivial solution is the only solution.

(b) The fact that $\det A = (1)(-4) - (-2)(2) = 0$ implies that A is singular. By Theorem 10, the corresponding system

$$x + 2y = 0$$
$$-2x - 4y = 0$$

must have non-trivial solutions. Note that by dividing by -2 we reduce the second equation to the first equation, $x + 2y = 0$. Letting $y = t$, we get $x = -2t$, and the solution of the system is given by

$$\{(-2t, t) \mid t \in \mathbb{R}\}$$

Clearly, if $t \neq 0$, the solution differs from $(0, 0)$; thus, it is a non-trivial solution of the given system.

Summary A square matrix A is called **invertible** or **non-singular** if there is a matrix B that multiplied by A in either order, gives the identity matrix. The matrix B is called the **inverse matrix** of A. A matrix is invertible if and only if its **determinant** is non-zero. By using an inverse matrix, we can solve $n \times n$ linear systems of equations in the case when the coefficient matrix is non-singular. A linear system is called **homogeneous** if all of its constant terms are zero. If the matrix of a homogeneous system is singular, then the system has a **non-trivial solution.**

9 Exercises

1. If A, B, and C are square matrices such that $ABC = CAB = I$, is it true that $C^{-1} = AB$?

2. If A, B, and C are square matrices such that $ABC = CBA = I$, is it true that $C^{-1} = AB$?

3. Assume that A is an invertible matrix. If $A^3 = I$, what is A^{-1}?

4. Find all values of a for which the matrix

$$\begin{bmatrix} a & a \\ a & 1 \end{bmatrix}$$

is non-singular.

5. Find all values of a for which the matrix

$$\begin{bmatrix} a & 2-a \\ a & 1 \end{bmatrix}$$

is invertible.

6. Let

$$A = \begin{bmatrix} a & b \\ c & d \end{bmatrix} \quad \text{and} \quad B = \begin{bmatrix} d/(ad-bc) & -b/(ad-bc) \\ -c/(ad-bc) & a/(ad-bc) \end{bmatrix}$$

Show that $BA = I$, thus completing the proof of Theorem 8.

7–16 ▪ Find the inverse of each matrix, or else say (and prove) that the matrix is not invertible.

7. $\begin{bmatrix} 0 & 1 \\ 2 & 3 \end{bmatrix}$

8. $\begin{bmatrix} 4 & -1 \\ 3 & 0 \end{bmatrix}$

9. $\begin{bmatrix} 3 & 4 \\ -6 & -8 \end{bmatrix}$

10. $\begin{bmatrix} 1 & 1 \\ 1 & 2 \end{bmatrix}$

11. $\begin{bmatrix} 0.1 & 0.2 \\ 0.4 & -0.2 \end{bmatrix}$

12. $\begin{bmatrix} -0.1 & 0 \\ 0.4 & -0.1 \end{bmatrix}$

13. $\begin{bmatrix} 1 & -2 & 3 \\ 0 & 1 & 0 \\ 0 & 1 & 0 \end{bmatrix}$

14. $\begin{bmatrix} 1 & 3 & -1 \\ -1 & 0 & 0 \\ 4 & 2 & 1 \end{bmatrix}$

15. $\begin{bmatrix} 1 & -2 & 3 \\ 4 & 1 & 0 \\ 0 & 1 & 1 \end{bmatrix}$

16. $\begin{bmatrix} 1 & 0 & 5 \\ -2 & 1 & 1 \\ 0 & 1 & 0 \end{bmatrix}$

17. Give an argument proving that the system

$$\begin{aligned} 2x - 6y &= 13 \\ 3x + 5y &= 7 \end{aligned}$$

has a unique solution. Find the solution using an inverse matrix.

18. Using an inverse matrix, find all solutions of the system

$$\begin{aligned} -3x + y &= 8 \\ 4x + 2y &= 5 \end{aligned}$$

Explain how you know that you have found all solutions.

19. Prove that the system

$$\begin{aligned} x + 3y - z &= 2 \\ -x + z &= 3 \\ 2x + 2y + z &= 1 \end{aligned}$$

has a unique solution. Using an inverse matrix, find the solution.

20. Prove that the system

$$\begin{aligned} 4x - z &= 3 \\ -3x + y + z &= 0 \\ -2y + z &= -1 \end{aligned}$$

has a unique solution. Using an inverse matrix, find the solution.

21. A square matrix A is called a *scalar* matrix if $A = \alpha I$, where α is a real number and I denotes the identity matrix. What is the inverse matrix of A?

22. A square matrix A is called *diagonal* if all of its off-diagonal terms are zero. State the condition(s) under which a 2×2 diagonal matrix has an inverse matrix, and find it.

23. A square matrix A is called *diagonal* if all of its off-diagonal terms are zero. State the condition(s) that guarantee that a 3×3 diagonal matrix has an inverse matrix, and find it.

24. Assume that A is an invertible matrix. Show that $(\alpha A)^{-1} = (1/\alpha)A^{-1}$.

25. Assume that A is an invertible 2×2 matrix. Show that the transposition commutes with the inverse, i.e., $(A^{-1})^t = (A^t)^{-1}$.

26. Show that for two invertible matrices A and B, $(B^{-1}A^{-1})(AB) = I$. Clearly identify the rationale for each step in your calculation.

27. Assume that you are trying to find the inverse matrix using Algorithm 2, and you end up with

$$\left[\begin{array}{cc|cc} 1 & 0 & 3 & 4 \\ 0 & 1 & 0 & 0 \end{array}\right]$$

How do you know that you made a mistake in your calculation?

28. Assume that θ is a real number, and define

$$A = \begin{bmatrix} \cos\theta & -\sin\theta \\ \sin\theta & \cos\theta \end{bmatrix}$$

Show that A is invertible and find A^{-1}.

29–34 ▪ In each case, solve the homogeneous system $A\mathbf{x} = \mathbf{0}$.

29. $\begin{bmatrix} 1 & 2 \\ 3 & 4 \end{bmatrix}$

30. $\begin{bmatrix} 2 & -4 \\ 1 & 5 \end{bmatrix}$

31. $\begin{bmatrix} -1 & 1 \\ -2 & 2 \end{bmatrix}$

32. $\begin{bmatrix} 2 & -1 \\ -1 & 2 \end{bmatrix}$

33. $\begin{bmatrix} 3 & 3 \\ 3 & 3 \end{bmatrix}$

34. $\begin{bmatrix} 7 & 0 \\ 0 & 0 \end{bmatrix}$

35. Assume that A is an invertible 2×2 matrix. Show that A^2 is invertible and find its inverse.

36. Assume that A is an invertible 2×2 matrix. Show that A^3 is invertible and find its inverse.

10 Linear Transformations

In this section, we start exploring **linear transformations,** a special kind of function whose domain and range consist of vectors in $\mathbb{R}^2$ and $\mathbb{R}^3$.

Introduction

Consider the 2×2 matrix

$$A = \begin{bmatrix} 1 & 2 \\ 0 & -3 \end{bmatrix}$$

and a vector $\mathbf{v} = \begin{bmatrix} 2 & 1 \end{bmatrix}$ in $\mathbb{R}^2$. (To save space, within a sentence we will use the row vector form.)

Using matrix multiplication, we transform the vector $\mathbf{v}$ into the vector $\mathbf{w}$ according to

$$\mathbf{w} = A \cdot \mathbf{v} = \begin{bmatrix} 1 & 2 \\ 0 & -3 \end{bmatrix} \begin{bmatrix} 2 \\ 1 \end{bmatrix} = \begin{bmatrix} 4 \\ -3 \end{bmatrix}$$

See Figure 10.1.

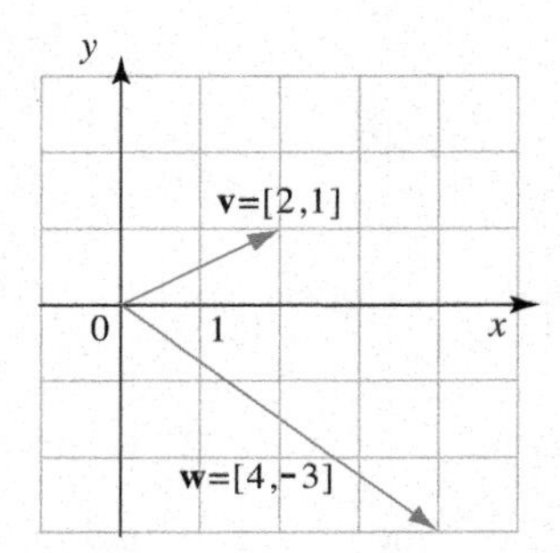

FIGURE 10.1
Transforming a vector

In this way, with the help of the matrix A, we build a function: to a vector $\mathbf{v}$ in $\mathbb{R}^2$ we assign a unique vector $\mathbf{w}$ in $\mathbb{R}^2$ given by $\mathbf{w} = A \cdot \mathbf{v}$. For example, if $\mathbf{v} = \begin{bmatrix} 0 & 1 \end{bmatrix}$, then

$$\mathbf{w} = A \cdot \mathbf{v} = \begin{bmatrix} 1 & 2 \\ 0 & -3 \end{bmatrix} \begin{bmatrix} 0 \\ 1 \end{bmatrix} = \begin{bmatrix} 2 \\ -3 \end{bmatrix}$$

and if $\mathbf{v} = \begin{bmatrix} -10 & 7 \end{bmatrix}$, then

$$\mathbf{w} = A \cdot \mathbf{v} = \begin{bmatrix} 1 & 2 \\ 0 & -3 \end{bmatrix} \begin{bmatrix} -10 \\ 7 \end{bmatrix} = \begin{bmatrix} 4 \\ -21 \end{bmatrix}$$

The vector $\mathbf{w}$ is called the *image* of the vector $\mathbf{v}$ under A. If $\mathbf{v} = \begin{bmatrix} x & y \end{bmatrix}$ is any vector in $\mathbb{R}^2$, then

$$\mathbf{w} = A \cdot \mathbf{v} = \begin{bmatrix} 1 & 2 \\ 0 & -3 \end{bmatrix} \begin{bmatrix} x \\ y \end{bmatrix} = \begin{bmatrix} x + 2y \\ -3y \end{bmatrix} \tag{10.1}$$

Formally, we define the function A by

$$A(\mathbf{v}) = A \cdot \mathbf{v} \tag{10.2}$$

where $\mathbf{v}$ is a vector in $\mathbb{R}^2$ and $\cdot$ is the symbol for matrix multiplication. (Because it will not be a cause of confusion, soon we will drop the multiplication sign $\cdot$ and just write $A(\mathbf{v}) = A\mathbf{v}$.)

The notation in (10.2) mirrors the usual functional notation (as in $f(x) = 7x$) that we use in calculus. We are dealing with matrices and prefer to use uppercase letters for the names of functions (the same letter for a function and the corresponding matrix) and $\mathbf{v}$ or $\mathbf{w}$ for the variable (because it is a vector). We often refer to the function A as a *transformation* or a *mapping*.

Note that the domain of A is $\mathbb{R}^2$ and its range is a subset of $\mathbb{R}^2$. Using (10.2), we can define a linear transformation in $\mathbb{R}^3$ or in $\mathbb{R}^n$ for any $n \geq 2$.

The linear transformation (10.1) can be written as $A(\mathbf{v}) = A \cdot \mathbf{v}$, i.e.,

$$A\left(\begin{bmatrix} x \\ y \end{bmatrix}\right) = \begin{bmatrix} 1 & 2 \\ 0 & -3 \end{bmatrix} \cdot \begin{bmatrix} x \\ y \end{bmatrix} = \begin{bmatrix} x + 2y \\ -3y \end{bmatrix}$$

To simplify notation, we drop the parentheses around the vector on the left side and write

$$A\begin{bmatrix} x \\ y \end{bmatrix} = \begin{bmatrix} 1 & 2 \\ 0 & -3 \end{bmatrix} \begin{bmatrix} x \\ y \end{bmatrix} = \begin{bmatrix} x + 2y \\ -3y \end{bmatrix} \tag{10.3}$$

Let's explore the properties of the transformation A. Take the vectors

$$\mathbf{v}_1 = \begin{bmatrix} 1 \\ 2 \end{bmatrix} \quad \text{and} \quad \mathbf{v}_2 = \begin{bmatrix} -2 \\ -1 \end{bmatrix}$$

in $\mathbb{R}^2$. Their sum is

$$\mathbf{v}_1 + \mathbf{v}_2 = \begin{bmatrix} -1 \\ 1 \end{bmatrix}$$

Using (10.3), we calculate the images of $\mathbf{v}_1$, $\mathbf{v}_2$, and $\mathbf{v}_1 + \mathbf{v}_2$ under A:

$$A(\mathbf{v}_1) = A\begin{bmatrix} 1 \\ 2 \end{bmatrix} = \begin{bmatrix} 1 & 2 \\ 0 & -3 \end{bmatrix} \begin{bmatrix} 1 \\ 2 \end{bmatrix} = \begin{bmatrix} 5 \\ -6 \end{bmatrix}$$

$$A(\mathbf{v}_2) = A\begin{bmatrix} -2 \\ -1 \end{bmatrix} = \begin{bmatrix} 1 & 2 \\ 0 & -3 \end{bmatrix} \begin{bmatrix} -2 \\ -1 \end{bmatrix} = \begin{bmatrix} -4 \\ 3 \end{bmatrix}$$

$$A(\mathbf{v}_1 + \mathbf{v}_2) = A\begin{bmatrix} -1 \\ 1 \end{bmatrix} = \begin{bmatrix} 1 & 2 \\ 0 & -3 \end{bmatrix} \begin{bmatrix} -1 \\ 1 \end{bmatrix} = \begin{bmatrix} 1 \\ -3 \end{bmatrix}$$

Note that

$$A(\mathbf{v}_1) + A(\mathbf{v}_2) = \begin{bmatrix} 5 \\ -6 \end{bmatrix} + \begin{bmatrix} -4 \\ 3 \end{bmatrix} = \begin{bmatrix} 1 \\ -3 \end{bmatrix}$$

i.e.,

$$A(\mathbf{v}_1 + \mathbf{v}_2) = A(\mathbf{v}_1) + A(\mathbf{v}_2) \tag{10.4}$$

We took two vectors, added them up, and then transformed the sum using A (that's the left side in (10.4)). As well, we transformed the two vectors using A first, and then added them up (the right side in (10.4)). The resulting vectors are equal.

The image of the vector $3\mathbf{v}_1$ is

$$A(3\mathbf{v}_1) = A\begin{bmatrix} 3 \\ 6 \end{bmatrix} = \begin{bmatrix} 1 & 2 \\ 0 & -3 \end{bmatrix} \begin{bmatrix} 3 \\ 6 \end{bmatrix} = \begin{bmatrix} 15 \\ -18 \end{bmatrix}$$

If we compute the image of $\mathbf{v}_1$ and then multiply by 3, we obtain

$$3A(\mathbf{v}_1) = 3A\begin{bmatrix} 1 \\ 2 \end{bmatrix} = 3\begin{bmatrix} 1 & 2 \\ 0 & -3 \end{bmatrix} \begin{bmatrix} 1 \\ 2 \end{bmatrix} = 3\begin{bmatrix} 5 \\ -6 \end{bmatrix} = \begin{bmatrix} 15 \\ -18 \end{bmatrix}$$

i.e.,

$$A(3\mathbf{v}_1) = 3A(\mathbf{v}_1) \tag{10.5}$$

Thus, it makes no difference whether we multiply the vector $\mathbf{v}_1$ by 3 first and then transform by A, or reverse the order (transform $\mathbf{v}_1$ by A first and then multiply by 3).

The properties (10.4) and (10.5) hold in general, for any vectors $\mathbf{v}_1$ and $\mathbf{v}_2$, for any real number, and for any matrix A (we will prove this soon). This is what makes the transformation A a *linear transformation.*

Sometimes we say that A is *compatible* with vector addition and multiplication by scalars. The precise meaning of the word "compatible" is given by (10.4) and (10.5) and by the next definition (Definition 27).

Linear Transformations

Consider the transformation

$$A(\mathbf{v}) = A \cdot \mathbf{v}$$

where A is a 2×2 matrix and $\mathbf{v}$ is a vector in $\mathbb{R}^2$. If

$$A = \begin{bmatrix} a & b \\ c & d \end{bmatrix} \quad \text{and} \quad \mathbf{v} = \begin{bmatrix} x \\ y \end{bmatrix}$$

then

$$A(\mathbf{v}) = \begin{bmatrix} a & b \\ c & d \end{bmatrix} \begin{bmatrix} x \\ y \end{bmatrix} = \begin{bmatrix} ax + by \\ cx + dy \end{bmatrix} \tag{10.6}$$

The function A is a function of two variables, x and y, whose range is a subset of $\mathbb{R}^2$. Note that the components of A are linear functions in x and y.

From the distributive property of matrix multiplication (property (c), Theorem 7 in Section 8) we conclude that

$$A(\mathbf{v}_1 + \mathbf{v}_2) = A \cdot (\mathbf{v}_1 + \mathbf{v}_2) = A \cdot \mathbf{v}_1 + A \cdot \mathbf{v}_2 = A(\mathbf{v}_1) + A(\mathbf{v}_2) \tag{10.7}$$

for $\mathbf{v}_1, \mathbf{v}_2 \in \mathbb{R}^2$. As well, by property (d) of the same theorem,

$$A(\alpha\mathbf{v}) = A \cdot (\alpha\mathbf{v}) = \alpha(A \cdot \mathbf{v}) = \alpha A(\mathbf{v}) \tag{10.8}$$

for $\alpha \in \mathbb{R}$ and $\mathbf{v} \in \mathbb{R}^2$. Any transformation A that satisfies (10.7) and (10.8) is called a *linear transformation.*

Definition 27 **Linear Transformation (Linear Mapping)**

A transformation A from $\mathbb{R}^n$ to $\mathbb{R}^n$ (where $n \geq 2$), that satisfies

(a) $A(\mathbf{v}_1 + \mathbf{v}_2) = A(\mathbf{v}_1) + A(\mathbf{v}_2)$ for $\mathbf{v}_1, \mathbf{v}_2 \in \mathbb{R}^n$ and

(b) $A(\alpha\mathbf{v}) = \alpha A(\mathbf{v})$ for $\mathbf{v} \in \mathbb{R}^n$ and $\alpha \in \mathbb{R}$

is called a *linear transformation* (or a *linear mapping*).

A linear transformation is said to "preserve" (or be compatible with) the operations of vector addition and multiplication by a scalar.

At the end of this section, we mention other interpretations of the function $A(\mathbf{v})$ introduced in (10.6). A linear map can be defined in more general terms, not necessarily for a square matrix (i.e., its domain and range do not have to be of the same dimension). We comment on this at the end of the section as well.

To learn about linear transformations and to gain experience in working with them without losing too much energy on messy calculations, we focus on 2×2 matrices.

Example 10.1 Linear Transformations

Describe the linear transformation $A(\mathbf{v}) = A\mathbf{v}$ defined by each matrix:

(a) identity matrix $A = I_2 = \begin{bmatrix} 1 & 0 \\ 0 & 1 \end{bmatrix}$

(b) scalar matrix $A = \begin{bmatrix} 3 & 0 \\ 0 & 3 \end{bmatrix}$ (where a *scalar* matrix is a multiple of the identity matrix)

(c) diagonal matrix $A = \begin{bmatrix} 2 & 0 \\ 0 & 1/5 \end{bmatrix}$ (which means that all off-diagonal entries in a *diagonal* matrix are zero)

▶ (a) Take any vector $\mathbf{v}$ in $\mathbb{R}^2$. Then

$$A(\mathbf{v}) = A \cdot \mathbf{v} = I_2 \cdot \mathbf{v} = \mathbf{v}$$

In coordinates,

$$A\begin{bmatrix} x \\ y \end{bmatrix} = \begin{bmatrix} 1 & 0 \\ 0 & 1 \end{bmatrix}\begin{bmatrix} x \\ y \end{bmatrix} = \begin{bmatrix} x \\ y \end{bmatrix}$$

Thus, the mapping that corresponds to the identity matrix leaves all vectors unchanged; it is called the *identity mapping* or the *identity transformation.*

(b) Writing $A = 3I_2$, we obtain

$$A(\mathbf{v}) = 3I_2 \cdot \mathbf{v} = 3\mathbf{v}$$

for any vector $\mathbf{v} \in \mathbb{R}^2$. Using coordinates, we write A as

$$A\begin{bmatrix} x \\ y \end{bmatrix} = \begin{bmatrix} 3 & 0 \\ 0 & 3 \end{bmatrix}\begin{bmatrix} x \\ y \end{bmatrix} = \begin{bmatrix} 3x \\ 3y \end{bmatrix}$$

The transformation A assigns, to each vector $\mathbf{v}$, the vector $3\mathbf{v}$; i.e., it stretches a vector by a factor of 3; see Figure 10.2a.

On the other hand, the transformation defined by the matrix

$$A = \begin{bmatrix} 1/2 & 0 \\ 0 & 1/2 \end{bmatrix}$$

contracts a vector: it keeps the direction of the vector, but makes it half as long; see Figure 10.2b.

The transformation given by the matrix

$$A = \begin{bmatrix} -4 & 0 \\ 0 & -4 \end{bmatrix}$$

expands a vector by a factor of 4 and then reflects it with respect to the origin; see Figure 10.2c.

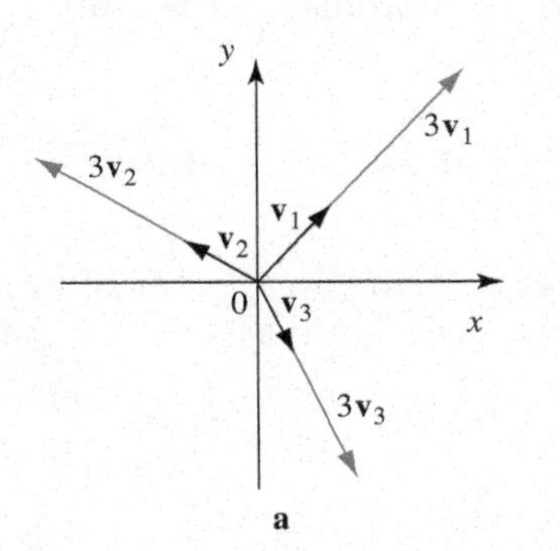

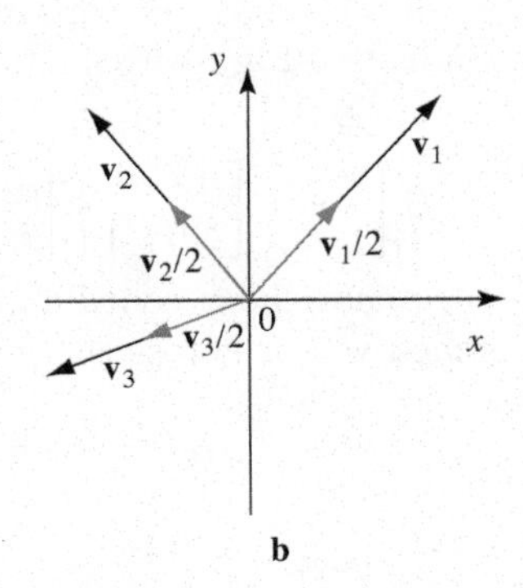

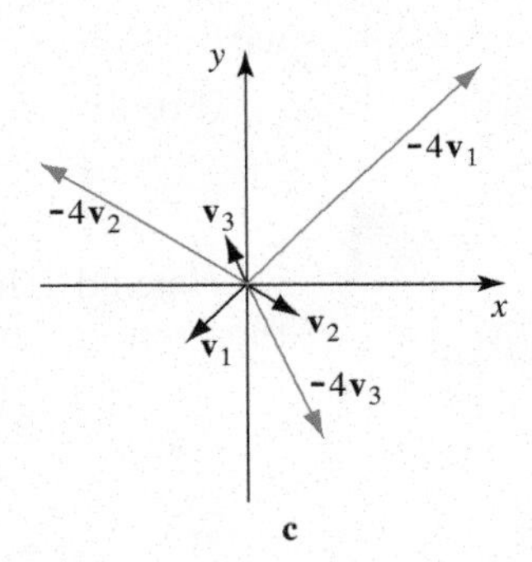

FIGURE 10.2
The transformations of vectors from Example 10.1(b)

(c) The transformation

$$A\begin{bmatrix} x \\ y \end{bmatrix} = \begin{bmatrix} 2 & 0 \\ 0 & 1/5 \end{bmatrix}\begin{bmatrix} x \\ y \end{bmatrix} = \begin{bmatrix} 2x \\ y/5 \end{bmatrix} \tag{10.9}$$

scales the coordinates of a vector differently: the first component is stretched by a factor of 2, and the second is compressed by a factor of 1/5. Note that, unlike the transformations in (a) and (b), this transformation changes the direction: a vector and its image under A are not, in general, parallel to each other. For instance,

$$A\begin{bmatrix} 1 \\ 1 \end{bmatrix} = \begin{bmatrix} 2 \\ 1/5 \end{bmatrix} \quad \text{and} \quad A\begin{bmatrix} -1 \\ -3 \end{bmatrix} = \begin{bmatrix} -2 \\ -3/5 \end{bmatrix}$$

See Figure 10.3.

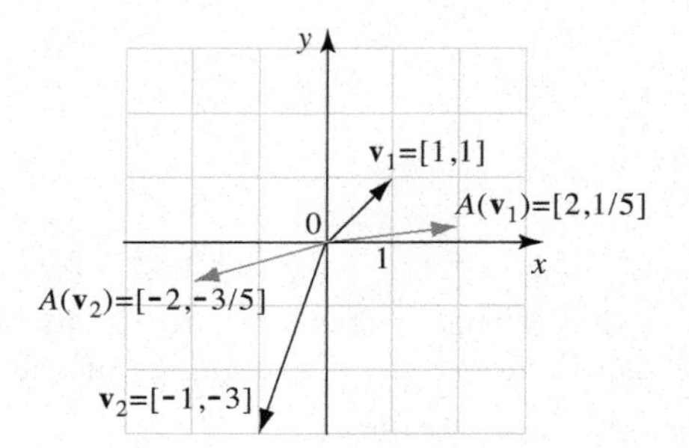

FIGURE 10.3
The transformation of Example 10.1(c)

The reason why we said "in general" is because there could be vectors whose direction is preserved (unchanged). For the given matrix, the directions of the coordinate axes are preserved: from (10.9) we compute

$$A\begin{bmatrix}1\\0\end{bmatrix}=\begin{bmatrix}2\\0\end{bmatrix} \quad \text{and} \quad A\begin{bmatrix}0\\1\end{bmatrix}=\begin{bmatrix}0\\1/5\end{bmatrix}$$

We will talk more about these special directions in Section 11.

Example 10.2 **Linear Transformations**

Describe the transformations defined by the matrices

$$A=\begin{bmatrix}1 & 0\\0 & -1\end{bmatrix} \quad \text{and} \quad B=\begin{bmatrix}-2 & 0\\0 & 1\end{bmatrix}$$

▶ The transformation defined by

$$A\begin{bmatrix}x\\y\end{bmatrix}=\begin{bmatrix}1 & 0\\0 & -1\end{bmatrix}\begin{bmatrix}x\\y\end{bmatrix}=\begin{bmatrix}x\\-y\end{bmatrix}$$

keeps the x-coordinate and changes the sign of the y-coordinate. It is a reflection with respect to the x-axis; see Figure 10.4a.

From

$$B\begin{bmatrix}x\\y\end{bmatrix}=\begin{bmatrix}-2 & 0\\0 & 1\end{bmatrix}\begin{bmatrix}x\\y\end{bmatrix}=\begin{bmatrix}-2x\\y\end{bmatrix}$$

we conclude that B stretches the x-coordinate of a vector by a factor of 2, and then reflects the stretched vector across the y-axis. See Figure 10.4b.

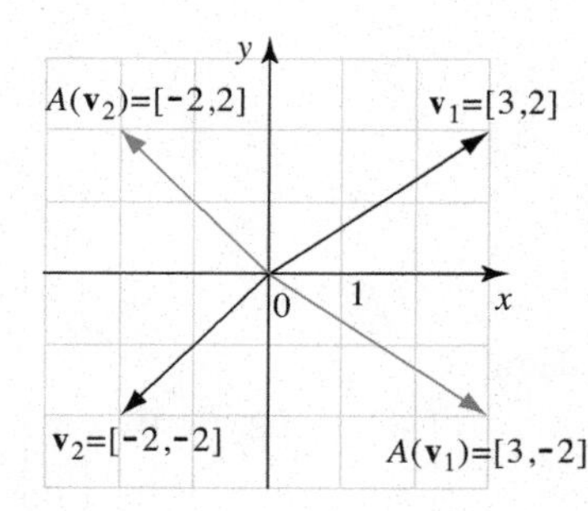

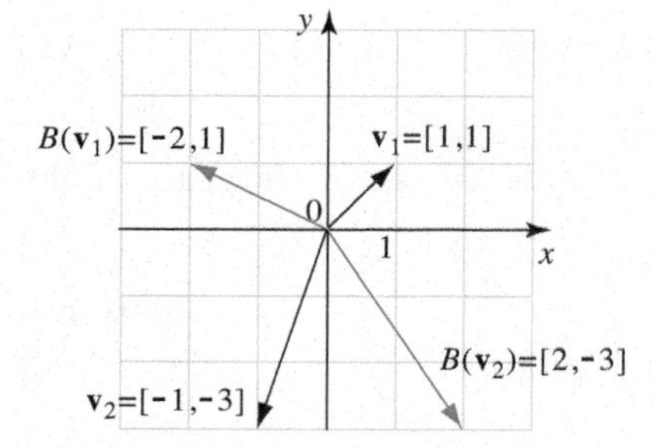

FIGURE 10.4
The transformations of Example 10.2

Example 10.3 **Rotation about the Origin**

Show that the transformation defined by

$$A=\begin{bmatrix}\cos\theta & -\sin\theta\\ \sin\theta & \cos\theta\end{bmatrix} \tag{10.10}$$

rotates a given vector about the origin by the angle θ.

▶ The image of a vector $\mathbf{v} = [x \quad y]$ is the vector

$$A(\mathbf{v}) = \begin{bmatrix} \cos\theta & -\sin\theta \\ \sin\theta & \cos\theta \end{bmatrix} \begin{bmatrix} x \\ y \end{bmatrix} = \begin{bmatrix} x\cos\theta - y\sin\theta \\ x\sin\theta + y\cos\theta \end{bmatrix} \tag{10.11}$$

Our plan is to show that the angle between any vector $\mathbf{v} \neq 0$ and its image $A(\mathbf{v})$ is equal to θ.

Recall that the angle α between two vectors $\mathbf{v}$ and $\mathbf{w}$ can be calculated from the formula

$$\cos\alpha = \frac{\mathbf{v}\cdot\mathbf{w}}{\|\mathbf{v}\|\,\|\mathbf{w}\|} \tag{10.12}$$

In our case, $\mathbf{w} = A(\mathbf{v})$. The dot product in the numerator is

$$\begin{aligned}
\mathbf{v}\cdot A(\mathbf{v}) &= \begin{bmatrix} x \\ y \end{bmatrix} \cdot \begin{bmatrix} x\cos\theta - y\sin\theta \\ x\sin\theta + y\cos\theta \end{bmatrix} \\
&= x(x\cos\theta - y\sin\theta) + y(x\sin\theta + y\cos\theta) \\
&= x^2\cos\theta - xy\sin\theta + xy\sin\theta + y^2\cos\theta \\
&= (x^2 + y^2)\cos\theta
\end{aligned}$$

The lengths in the denominator are computed to be

$$\|\mathbf{v}\| = \sqrt{x^2 + y^2}$$

and

$$\begin{aligned}
\|A(\mathbf{v})\| &= \left((x\cos\theta - y\sin\theta)^2 + (x\sin\theta + y\cos\theta)^2\right)^{1/2} \\
&= \big(x^2\cos^2\theta - 2xy\sin\theta\cos\theta + y^2\sin^2\theta + x^2\sin^2\theta + 2xy\sin\theta\cos\theta \\
&\qquad + y^2\cos^2\theta\big)^{1/2} \\
&= \left(x^2(\cos^2\theta + \sin^2\theta) + y^2(\sin^2\theta + \cos^2\theta)\right)^{1/2} \\
&= \left(x^2 + y^2\right)^{1/2}
\end{aligned}$$

Substituting into (10.12) we obtain

$$\cos\alpha = \frac{(x^2 + y^2)\cos\theta}{\sqrt{x^2 + y^2}\,\sqrt{x^2 + y^2}} = \cos\theta \tag{10.13}$$

Thus, $\theta = \alpha$ or $\theta = 2\pi - \alpha$. In either case, the angle of rotation is θ (recall Definition 12 in Section 3, which says that the angle between two vectors is the smaller of the two angles made by their directions).

In Exercise 18 we suggest an alternative way of proving this fact; as well, we show that (10.10) represents a counterclockwise rotation if $\theta > 0$. ▲

Looking at the intermediate results in Example 10.3, we note that $\|\mathbf{v}\| = \|A(\mathbf{v})\|$. This means that the transformation A preserves the length of a vector.

Substituting $\theta = \pi/4$ into (10.11), we obtain the formula

$$A\begin{bmatrix} x \\ y \end{bmatrix} = \begin{bmatrix} \cos(\pi/4) & -\sin(\pi/4) \\ \sin(\pi/4) & \cos(\pi/4) \end{bmatrix} \begin{bmatrix} x \\ y \end{bmatrix} = \begin{bmatrix} \sqrt{2}x/2 - \sqrt{2}y/2 \\ \sqrt{2}x/2 + \sqrt{2}y/2 \end{bmatrix}$$

for a rotation by $\pi/4$ radians. For instance,

$$A\begin{bmatrix} 1 \\ 0 \end{bmatrix} = \begin{bmatrix} \sqrt{2}/2 \\ \sqrt{2}/2 \end{bmatrix}$$

$$A\begin{bmatrix} 1 \\ 1 \end{bmatrix} = \begin{bmatrix} \sqrt{2}/2 - \sqrt{2}/2 \\ \sqrt{2}/2 + \sqrt{2}/2 \end{bmatrix} = \begin{bmatrix} 0 \\ \sqrt{2} \end{bmatrix}$$

$$A\begin{bmatrix} -1 \\ 0 \end{bmatrix} = \begin{bmatrix} -\sqrt{2}/2 \\ -\sqrt{2}/2 \end{bmatrix}$$

and so on. This is a counterclockwise rotation; see Figure 10.5.

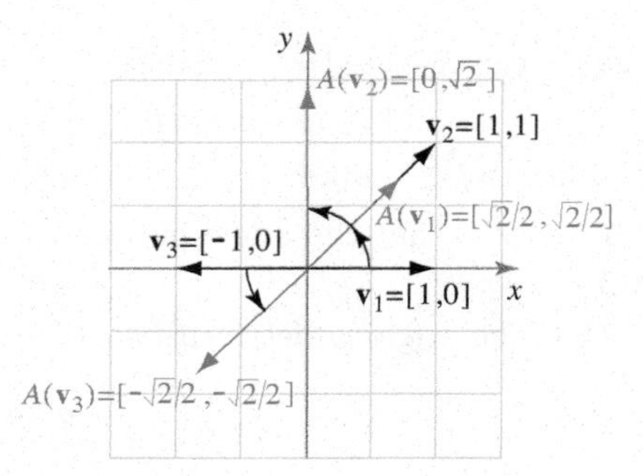

FIGURE 10.5

Rotation is an example of a linear transformation

Example 10.4 Linear Transformation in $\mathbb{R}^3$

Interpret geometrically the transformation in $\mathbb{R}^3$ defined by the matrix

$$A = \begin{bmatrix} \cos\theta & -\sin\theta & 0 \\ \sin\theta & \cos\theta & 0 \\ 0 & 0 & 1 \end{bmatrix}$$

▶ In the upper left corner of A we recognize the matrix of rotation in the xy-plane that was introduced in (10.10) in Example 10.3. Take a vector $\mathbf{v} = [x \quad y \quad z]$ in $\mathbb{R}^3$. Its projection onto the xy-plane is the vector $\mathbf{v} = [x \quad y \quad 0]$. The calculation

$$A\begin{bmatrix} x \\ y \\ 0 \end{bmatrix} = \begin{bmatrix} \cos\theta & -\sin\theta & 0 \\ \sin\theta & \cos\theta & 0 \\ 0 & 0 & 1 \end{bmatrix}\begin{bmatrix} x \\ y \\ 0 \end{bmatrix} = \begin{bmatrix} x\cos\theta - y\sin\theta \\ x\sin\theta + y\cos\theta \\ 0 \end{bmatrix}$$

tells us that this projection remains in the xy-plane (since its z-coordinate is zero), and that it is subjected to a rotation by the angle θ; see Figure 10.6. From

$$A\begin{bmatrix} 0 \\ 0 \\ z \end{bmatrix} = \begin{bmatrix} \cos\theta & -\sin\theta & 0 \\ \sin\theta & \cos\theta & 0 \\ 0 & 0 & 1 \end{bmatrix}\begin{bmatrix} 0 \\ 0 \\ z \end{bmatrix} = \begin{bmatrix} 0 \\ 0 \\ z \end{bmatrix}$$

we conclude that the z-coordinate of a vector does not change. Thus, the matrix A represents a rotation about the z-axis by the angle θ. ▲

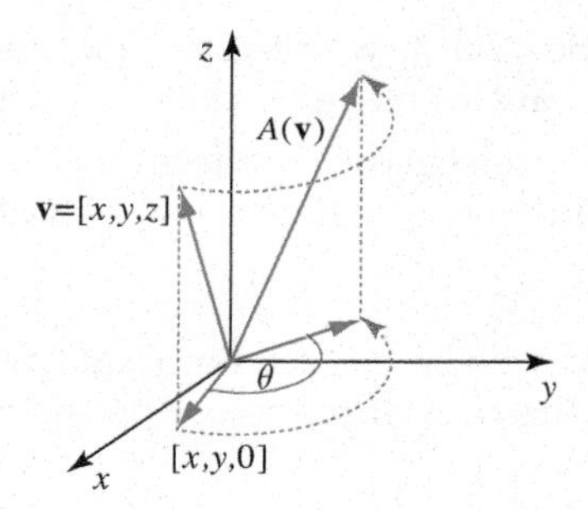

FIGURE 10.6

Rotation about the z-axis

Example 10.5 Composition of Transformations

Consider the linear transformations

$$A\begin{bmatrix} x \\ y \end{bmatrix} = \begin{bmatrix} 3 & 4 \\ 5 & 6 \end{bmatrix}\begin{bmatrix} x \\ y \end{bmatrix} = \begin{bmatrix} 3x+4y \\ 5x+6y \end{bmatrix}$$

and

$$B\begin{bmatrix} x \\ y \end{bmatrix} = \begin{bmatrix} -1 & 0 \\ 2 & 5 \end{bmatrix}\begin{bmatrix} x \\ y \end{bmatrix} = \begin{bmatrix} -x \\ 2x+5y \end{bmatrix}$$

Since A and B are functions, we can form their composition:

$$(B \circ A)\begin{bmatrix} x \\ y \end{bmatrix} = B\left(A\begin{bmatrix} x \\ y \end{bmatrix}\right)$$
$$= B\begin{bmatrix} 3x+4y \\ 5x+6y \end{bmatrix}$$
$$= \begin{bmatrix} -1 & 0 \\ 2 & 5 \end{bmatrix}\begin{bmatrix} 3x+4y \\ 5x+6y \end{bmatrix}$$
$$= \begin{bmatrix} -(3x+4y) \\ 2(3x+4y)+5(5x+6y) \end{bmatrix}$$
$$= \begin{bmatrix} -3x-4y \\ 31x+38y \end{bmatrix}$$

Now we calculate the product of the two matrices

$$C = BA = \begin{bmatrix} -1 & 0 \\ 2 & 5 \end{bmatrix}\begin{bmatrix} 3 & 4 \\ 5 & 6 \end{bmatrix} = \begin{bmatrix} -3 & -4 \\ 31 & 38 \end{bmatrix}$$

and the corresponding transformation

$$C\begin{bmatrix} x \\ y \end{bmatrix} = \begin{bmatrix} -3 & -4 \\ 31 & 38 \end{bmatrix}\begin{bmatrix} x \\ y \end{bmatrix} = \begin{bmatrix} -3x-4y \\ 31x+38y \end{bmatrix}$$

So, in this case, the composition of transformations $B \circ A$ is the linear transformation defined by the product BA of the matrices B and A (in that order; multiplication of matrices is not commutative, and neither is the composition of functions).

What we have discovered is true in general: the matrix representing the composition $B \circ A$ of two linear transformations is the product of the matrices B and A representing the two transformations (see Exercise 26). ▲

Remarks We defined a linear transformation as a function from $\mathbb{R}^2$ to $\mathbb{R}^2$ (or from $\mathbb{R}^3$ to $\mathbb{R}^3$, or from $\mathbb{R}^n$ to $\mathbb{R}^n$, $n \geq 2$; since the dimension is not relevant at the moment, we stick with $n = 2$). Using functional notation, we write $A: \mathbb{R}^2 \to \mathbb{R}^2$. Recall that $\mathbb{R}^2$ can be interpreted either as a set of points or as a set of vectors.

When we interpret $A: \mathbb{R}^2 \to \mathbb{R}^2$ as an assignment of a vector in $\mathbb{R}^2$ to a *point* in $\mathbb{R}^2$, we obtain a *linear vector field.* (Thus, the domain of a linear vector field is the set of all points in $\mathbb{R}^2$.) We interpret the assignment

$$A(x, y) = \begin{bmatrix} 2 & -1 \\ 1 & 1 \end{bmatrix}\begin{bmatrix} x \\ y \end{bmatrix} = \begin{bmatrix} 2x-y \\ x+y \end{bmatrix}$$

as a vector whose tail is located at (x, y). For instance,

$$A(1,0) = \begin{bmatrix} 2 \\ 1 \end{bmatrix}, \quad A(-1,-1) = \begin{bmatrix} -1 \\ -2 \end{bmatrix}, \quad \text{and} \quad A(0,2) = \begin{bmatrix} -2 \\ 2 \end{bmatrix}$$

are shown as vectors in Figure 10.7.

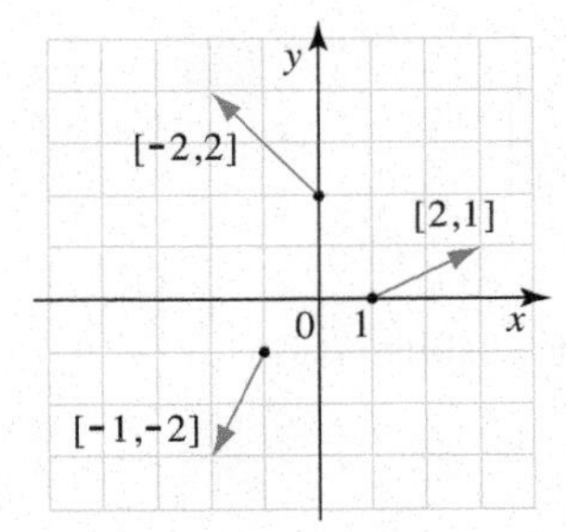

FIGURE 10.7
Linear vector field

We obtain yet another interpretation if we think of both the domain and the range of $A\colon \mathbb{R}^2 \to \mathbb{R}^2$ as sets of points: the function

$$A(x, y) = \begin{bmatrix} 2 & -1 \\ 1 & 1 \end{bmatrix} \begin{bmatrix} x \\ y \end{bmatrix} = (2x - y, x + y)$$

is an example of a function of two variables whose range is a subset of $\mathbb{R}^2$.

Depending on the context, these different appearances of essentially the same function become quite important.

It is possible to define linear maps using non-square matrices. For instance

$$A \begin{bmatrix} x \\ y \end{bmatrix} = \begin{bmatrix} 1 & -2 \\ -3 & 0 \\ -5 & 6 \end{bmatrix} \begin{bmatrix} x \\ y \end{bmatrix} = \begin{bmatrix} x - 2y \\ -3x \\ -5x + 6y \end{bmatrix}$$

is a transformation from $\mathbb{R}^2$ into $\mathbb{R}^3$. Although they are important, we do not study such transformations in this module.

Summary We can use matrices to build functions whose independent and dependent variables are vectors. These functions are called **linear transformations** or **linear mappings.** To find the value of a linear transformation at a vector, we multiply the matrix of the transformation by the vector. Linear transformations **preserve** the addition of vectors and the multiplication of a vector by a scalar. The **composition** of linear transformations corresponds to the **product** of their matrices.

10 Exercises

1. If A is a linear transformation such that

$$A(\mathbf{v}_1) = \begin{bmatrix} 3 \\ -9 \end{bmatrix} \quad \text{and} \quad A(\mathbf{v}_2) = \begin{bmatrix} 10 \\ 1 \end{bmatrix}$$

find $A(9\mathbf{v}_1)$, $A(-\mathbf{v}_2)$, and $A(2\mathbf{v}_1 - 3\mathbf{v}_2)$.

2. It is known that B is a linear transformation and

$$B(\mathbf{v}_1) = \begin{bmatrix} 0 \\ -4 \end{bmatrix} \quad \text{and} \quad B(\mathbf{v}_2) = \begin{bmatrix} 4 \\ 1 \end{bmatrix}$$

Find $B(-2\mathbf{v}_2)$, $B(\mathbf{v}_1 + \mathbf{v}_2)$, and $B(-10\mathbf{v}_1 + 3\mathbf{v}_2)$.

3. Assume that A is a linear transformation, and $A(\mathbf{v}) = [3 \quad -4]$. Is it true that $A(-\mathbf{v}) = [-3 \quad 4]$?

4. Explain why there is no linear transformation B such that

$$B \begin{bmatrix} 4 \\ -5 \end{bmatrix} = \begin{bmatrix} 2 \\ -3 \end{bmatrix} \quad \text{and} \quad B \begin{bmatrix} -4 \\ 5 \end{bmatrix} = \begin{bmatrix} -2 \\ -3 \end{bmatrix}$$

5. By using a counterexample, show that the transformation

$$A \begin{bmatrix} x \\ y \end{bmatrix} = \begin{bmatrix} x + 2y \\ 3x - y + 4 \end{bmatrix}$$

is not linear.

6. Give a counterexample to show that the transformation

$$A \begin{bmatrix} x \\ y \end{bmatrix} = \begin{bmatrix} xy \\ x - y \end{bmatrix}$$

is not linear.

7. Consider the transformation $A(\mathbf{v}) = A\mathbf{v}$. Knowing that

$$A = \begin{bmatrix} 3 & -2 \\ 4 & 0 \end{bmatrix}, \quad \mathbf{v} = \begin{bmatrix} 10 \\ 2 \end{bmatrix}, \quad \text{and} \quad \mathbf{w} = \begin{bmatrix} -1 \\ -6 \end{bmatrix},$$

find $A(\mathbf{v})$, $A(\mathbf{w})$, and $A(2\mathbf{v} + 4\mathbf{w})$.

8. Consider the transformation $A(\mathbf{v}) = A\mathbf{v}$. Given that

$$A = \begin{bmatrix} -1 & 2 \\ 5 & 6 \end{bmatrix}, \quad \mathbf{v} = \begin{bmatrix} 0 \\ 4 \end{bmatrix}, \quad \text{and} \quad \mathbf{w} = \begin{bmatrix} 3 \\ -5 \end{bmatrix},$$

find $A(\mathbf{v})$, $A(\mathbf{w})$, and $A(\mathbf{v} - 5\mathbf{w})$.

9. Consider the transformation $A(\mathbf{v}) = A\mathbf{v}$. Find a matrix A such that

$$A\begin{bmatrix} 1 \\ 0 \end{bmatrix} = \begin{bmatrix} 2 \\ -7 \end{bmatrix} \quad \text{and} \quad A\begin{bmatrix} 0 \\ 1 \end{bmatrix} = \begin{bmatrix} -2 \\ 3 \end{bmatrix}$$

10. Consider the transformation $A(\mathbf{v}) = A\mathbf{v}$. Find the matrix A if it is known that

$$A\begin{bmatrix} 2 \\ 0 \end{bmatrix} = \begin{bmatrix} -8 \\ 4 \end{bmatrix} \quad \text{and} \quad A\begin{bmatrix} 0 \\ 3 \end{bmatrix} = \begin{bmatrix} 6 \\ 9 \end{bmatrix}$$

11–16 ▪ Find a formula for the transformation defined by each matrix A (i.e., show how the coordinates of a vector are transformed). Give a geometric interpretation of each transformation.

11. $A = \begin{bmatrix} 0 & 0 \\ 0 & 1 \end{bmatrix}$

12. $A = \begin{bmatrix} 2 & 0 \\ 0 & 0 \end{bmatrix}$

13. $A = \begin{bmatrix} -1 & 0 \\ 0 & 1 \end{bmatrix}$

14. $A = \begin{bmatrix} -1 & 0 \\ 0 & -1 \end{bmatrix}$

15. $A = \begin{bmatrix} -2 & 0 \\ 0 & 1 \end{bmatrix}$

16. $A = \begin{bmatrix} 1/2 & 0 \\ 0 & 1/4 \end{bmatrix}$

17. How would you change the matrix A in Example 10.3 to make the rotation clockwise?

18. An alternative way of proving the claim in Example 10.3 is to consider a vector in polar form: write $\mathbf{v} = [r\cos\alpha \quad r\sin\alpha]$ and show that $A(\mathbf{v})$ is of the same form as A, with the angle θ replaced by $\theta + \alpha$. Conclude that the rotation is counterclockwise if $\theta > 0$.

19. Give a geometric interpretation of the transformation in $\mathbb{R}^2$ given by the matrix

$$A = \begin{bmatrix} 1 & 1 \\ 1 & 1 \end{bmatrix}$$

20–25 ▪ Find the formula for the linear transformation $A(\mathbf{v}) = A\mathbf{v}$ (i.e., find the matrix A) for each geometric transformation.

20. A stretches a vector by a factor of 3 and then reflects it across the x-axis.

21. A contracts a vector by a factor of $1/4$ and then reflects with respect to the origin.

22. A reflects a vector across the line $y = x$.

23. A reflects a vector across the line $y = -x$.

24. A rotates a vector by $3\pi/4$ radians counterclockwise about the origin.

25. A rotates a vector by $5\pi/6$ radians counterclockwise about the origin.

26. Assume that A and B are 2×2 matrices, and let $C = BA$. Show that the composition $B \circ A$ of the transformations defined by B and A is the linear transformation defined by the matrix C.

27. Let $A(\mathbf{v})$ be a linear transformation defined by a non-singular matrix A, i.e., $A(\mathbf{v}) = A\mathbf{v}$. Find the matrix corresponding to the inverse transformation of $A(\mathbf{v})$.

28. Find A^{-1} for the rotation matrix A given by (10.10) in Example 10.3. What geometric transformation is represented by A^{-1}?

29. Assume that A and B are linear transformations. Using Definition 27, show that the composition $A \circ B$ is also a linear transformation.

30. Given that

$$A = \begin{bmatrix} 1 & 2 & 1 \\ 0 & 0 & 2 \\ 1 & -2 & 4 \end{bmatrix}, \quad \mathbf{v} = \begin{bmatrix} -1 \\ 0 \\ 2 \end{bmatrix}, \quad \text{and} \quad \mathbf{w} = \begin{bmatrix} 0 \\ 4 \\ -6 \end{bmatrix},$$

find $A(\mathbf{v})$, $A(\mathbf{w})$, and $A(2\mathbf{v} + 4\mathbf{w})$.

31. Given that

$$A = \begin{bmatrix} 1 & 0 & -4 \\ 3 & 0 & 1 \\ 0 & -2 & 1 \end{bmatrix}, \quad \mathbf{v} = \begin{bmatrix} -1 \\ 1 \\ 1 \end{bmatrix}, \quad \text{and} \quad \mathbf{w} = \begin{bmatrix} 0 \\ 0 \\ 2 \end{bmatrix},$$

find $A(5\mathbf{v})$, $A(4\mathbf{w})$, and $A(5\mathbf{v} + 4\mathbf{w})$.

32. Give a geometric interpretation of the transformation in $\mathbb{R}^3$ given by the matrix

$$A = \begin{bmatrix} 1 & 0 & 0 \\ 0 & 1 & 0 \\ 0 & 0 & 0 \end{bmatrix}$$

33. Give a geometric interpretation of the transformation in $\mathbb{R}^3$ given by the matrix

$$A = \begin{bmatrix} 1 & 0 & 0 \\ 0 & 1 & 0 \\ 0 & 0 & -1 \end{bmatrix}$$

34. Find the formula (i.e., find the matrix A) for the linear transformation $A(\mathbf{v}) = A\mathbf{v}$ in $\mathbb{R}^3$ that projects a vector onto the xz-plane.

35. Find the formula (i.e., find the matrix A) for the linear transformation $A(\mathbf{v}) = A\mathbf{v}$ in $\mathbb{R}^3$ that reflects a vector with respect to the xz-plane.

11 Eigenvalues and Eigenvectors

In Example 10.1 in Section 10 we noticed that there are vectors whose direction is preserved by a linear transformation; i.e., the image $A(\mathbf{v})$ of a vector $\mathbf{v}$ under a transformation A is parallel to $\mathbf{v}$. We now study this situation in more detail.

Consider the transformation A given by

$$A\begin{bmatrix} x \\ y \end{bmatrix} = \begin{bmatrix} 1 & 3 \\ 4 & 2 \end{bmatrix}\begin{bmatrix} x \\ y \end{bmatrix}$$

In general, A changes the direction of the vector it acts upon. For instance, the vector $\mathbf{v} = [1 \quad 4]$ and its image

$$A\begin{bmatrix} 1 \\ 4 \end{bmatrix} = \begin{bmatrix} 1 & 3 \\ 4 & 2 \end{bmatrix}\begin{bmatrix} 1 \\ 4 \end{bmatrix} = \begin{bmatrix} 13 \\ 12 \end{bmatrix}$$

point in different directions. The fact that

$$A\begin{bmatrix} 0 \\ 1 \end{bmatrix} = \begin{bmatrix} 1 & 3 \\ 4 & 2 \end{bmatrix}\begin{bmatrix} 0 \\ 1 \end{bmatrix} = \begin{bmatrix} 3 \\ 2 \end{bmatrix}$$

shows that the vertical direction (the direction of the y-axis) is transformed under A into the direction of a line with a slope of $2/3$.

On the other hand, the image of the vector $\mathbf{v} = [1 \quad -1]$

$$A\begin{bmatrix} 1 \\ -1 \end{bmatrix} = \begin{bmatrix} 1 & 3 \\ 4 & 2 \end{bmatrix}\begin{bmatrix} 1 \\ -1 \end{bmatrix} = \begin{bmatrix} -2 \\ 2 \end{bmatrix}$$

is parallel to $\mathbf{v}$; i.e.,

$$A\begin{bmatrix} 1 \\ -1 \end{bmatrix} = (-2)\begin{bmatrix} 1 \\ -1 \end{bmatrix}$$

How do we find these special vectors (special directions)? (Soon, we will give them a name.) How many special directions does a linear transformation have?

The linear transformation defined by the matrix

$$A = \begin{bmatrix} 3 & 0 \\ 0 & 3 \end{bmatrix}$$

maps every vector onto its multiple (see Example 10.1(b)) and therefore preserves every direction. Rotation in a plane (Example 10.3) does not preserve any direction if the angle of rotation θ satisfies $\theta \neq k\pi$, where $k = 0, \pm 1, \pm 2, \ldots$.

Thus, a linear transformation might have no special directions, or some or all directions could be special (in the sense of being preserved under the transformation).

Eigenvalues and Eigenvectors

We now give a precise meaning to the ideas discussed in the introduction.

Definition 28 **Eigenvalue and Eigenvector**

Assume that A is a square matrix. A non-zero vector $\mathbf{v}$ for which there exists a real number λ such that

$$A\mathbf{v} = \lambda\mathbf{v} \tag{11.1}$$

is called an *eigenvector* of A. The number λ is called an *eigenvalue* of A. ▲

To emphasize the bond between an eigenvalue and an eigenvector given by (11.1), we often say "an eigenvector $\mathbf{v}$ corresponding to the eigenvalue λ" or "an eigenvalue λ with a corresponding eigenvector $\mathbf{v}$."

Example 11.1 Eigenvectors and Eigenvalues

The equation

$$A\begin{bmatrix}1\\-1\end{bmatrix}=\begin{bmatrix}-2\\2\end{bmatrix}=(-2)\begin{bmatrix}1\\-1\end{bmatrix}$$

implies that $\lambda=-2$ is an eigenvalue of A and $\mathbf{v}=[1 \quad -1]$ is the corresponding eigenvector. From

$$\begin{bmatrix}3&-2\\0&5\end{bmatrix}\begin{bmatrix}4\\0\end{bmatrix}=\begin{bmatrix}12\\0\end{bmatrix}=3\begin{bmatrix}4\\0\end{bmatrix}$$

we conclude that $\mathbf{v}=[4 \quad 0]$ is an eigenvector for the matrix

$$B=\begin{bmatrix}3&-2\\0&5\end{bmatrix}$$

corresponding to the eigenvalue $\lambda=3$. The vector $[2 \quad -1]$ is not an eigenvector of B, since the vector

$$B\mathbf{v}=\begin{bmatrix}3&-2\\0&5\end{bmatrix}\begin{bmatrix}2\\-1\end{bmatrix}=\begin{bmatrix}8\\-5\end{bmatrix}$$

is not parallel to $[2 \quad -1]$.

Recall that a square matrix A defines a linear transformation (also denoted by A) by matrix multiplication: $A(\mathbf{v})=A\mathbf{v}$. Thus, if $\mathbf{v}$ is an eigenvector of the matrix A with the corresponding eigenvalue λ, then $A(\mathbf{v})=A\mathbf{v}=\lambda\mathbf{v}$. With the equation $A(\mathbf{v})=\lambda\mathbf{v}$ in mind, we say that λ is an eigenvalue of the *transformation* A, and $\mathbf{v}$ is the corresponding eigenvector. (Very often we blur the distinction between a matrix and the corresponding linear transformation.)

The relation $A(\mathbf{v})=\lambda\mathbf{v}$ says that the transformation A maps the eigenvector $\mathbf{v}$ onto its scalar multiple (note that λ could be zero; see Example 11.5).

The zero vector $\mathbf{0}$ satisfies $A(\mathbf{0})=\mathbf{0}=\lambda\cdot\mathbf{0}$ for any linear transformation and any real number λ. To exclude this trivial situation, Definition 28 requires that an eigenvector be a non-zero vector.

If $\mathbf{v}$ is an eigenvector of A, then any multiple $c\mathbf{v}$ of $\mathbf{v}$ ($c\neq 0$) is also an eigenvector with the same eigenvalue, since

$$A(c\mathbf{v})=cA(\mathbf{v})=c\lambda\mathbf{v}=\lambda(c\mathbf{v})$$

In other words, if $\mathbf{v}$ is an eigenvalue, then the whole *line* through the origin whose direction is given by $\mathbf{v}$ is preserved by the transformation A. (The line is preserved as a set, not point by point; i.e., individual points on the line (which we think of as tips of vectors) do get moved by A, but to other points on the line.) Thus, to an eigenvalue we can assign not only a vector but a whole line through the origin.

How do we find eigenvalues and eigenvectors?

To avoid complicated calculations, we work with 2×2 matrices.

Example 11.2 Finding Eigenvalues and Eigenvectors

Consider the matrix

$$A=\begin{bmatrix}1&3\\4&2\end{bmatrix}$$

from the introduction. We are looking for a non-zero vector $\mathbf{v}=[x \quad y]$ and a real number λ such that

$$A\mathbf{v}=\lambda\mathbf{v} \tag{11.2}$$

Writing (11.2) in coordinates, we obtain

$$\begin{bmatrix} 1 & 3 \\ 4 & 2 \end{bmatrix} \begin{bmatrix} x \\ y \end{bmatrix} = \lambda \begin{bmatrix} x \\ y \end{bmatrix}$$

$$\begin{bmatrix} x + 3y \\ 4x + 2y \end{bmatrix} = \begin{bmatrix} \lambda x \\ \lambda y \end{bmatrix}$$

The corresponding linear system

$$\begin{aligned} x + 3y &= \lambda x \\ 4x + 2y &= \lambda y \end{aligned}$$

reduces to

$$\begin{aligned} (1 - \lambda)x + 3y &= 0 \\ 4x + (2 - \lambda)y &= 0 \end{aligned} \qquad (11.3)$$

The system (11.3) is a homogeneous linear system in two variables, x and y. By Theorem 10 in Section 9, this system has a non-trivial (i.e., non-zero) solution for x and y if and only if the determinant of the coefficient matrix

$$\begin{bmatrix} 1 - \lambda & 3 \\ 4 & 2 - \lambda \end{bmatrix}$$

is zero. Thus,

$$\begin{aligned} \begin{vmatrix} 1 - \lambda & 3 \\ 4 & 2 - \lambda \end{vmatrix} &= 0 \\ (1 - \lambda)(2 - \lambda) - 12 &= 0 \\ \lambda^2 - 3\lambda - 10 &= 0 \\ (\lambda - 5)(\lambda + 2) &= 0 \end{aligned}$$

Thus, there are two eigenvalues, $\lambda_1 = 5$ and $\lambda_2 = -2$.

For each eigenvalue, we find a corresponding eigenvector. When $\lambda_1 = 5$, the equation $A\mathbf{v} = \lambda_1 \mathbf{v}$ implies

$$\begin{bmatrix} 1 & 3 \\ 4 & 2 \end{bmatrix} \begin{bmatrix} x \\ y \end{bmatrix} = 5 \begin{bmatrix} x \\ y \end{bmatrix}$$

where the variables are the coordinates of the eigenvector $\mathbf{v} = [x \quad y]$. Writing this matrix equation in coordinates, we obtain the system

$$\begin{aligned} x + 3y &= 5x \\ 4x + 2y &= 5y \end{aligned}$$

and, after simplifying,

$$\begin{aligned} -4x + 3y &= 0 \\ 4x - 3y &= 0 \end{aligned}$$

The two equations are identical. To solve $4x - 3y = 0$, we use a parameter: let $y = t$; then $4x - 3t = 0$ and $x = 3t/4$. We conclude that any vector of the form

$$\mathbf{v} = \begin{bmatrix} x \\ y \end{bmatrix} = \begin{bmatrix} 3t/4 \\ t \end{bmatrix} = t \begin{bmatrix} 3/4 \\ 1 \end{bmatrix} \qquad (11.4)$$

where $t \neq 0$, is an eigenvector for $\lambda_1 = 5$.

All eigenvectors corresponding to $\lambda_1 = 5$ lie on the line through the origin whose direction is given by the vector $[3/4 \quad 1]$. For example, when $t = 4$, the formula (11.4) gives the vector $\mathbf{v} = [3 \quad 4]$. To check:

$$\begin{aligned} A\mathbf{v} &= A \begin{bmatrix} 3 \\ 4 \end{bmatrix} = \begin{bmatrix} 1 & 3 \\ 4 & 2 \end{bmatrix} \begin{bmatrix} 3 \\ 4 \end{bmatrix} = \begin{bmatrix} 15 \\ 20 \end{bmatrix} \\ 5\mathbf{v} &= 5 \begin{bmatrix} 3 \\ 4 \end{bmatrix} = \begin{bmatrix} 15 \\ 20 \end{bmatrix} \end{aligned}$$

so, indeed, the equation $A(\mathbf{v}) = 5\mathbf{v}$ holds with $\mathbf{v} = [3 \quad 4]$.

We repeat this routine to find an eigenvector for the eigenvalue $\lambda_2 = -2$. The matrix equation

$$\begin{bmatrix} 1 & 3 \\ 4 & 2 \end{bmatrix} \begin{bmatrix} x \\ y \end{bmatrix} = -2 \begin{bmatrix} x \\ y \end{bmatrix}$$

yields the linear system

$$\begin{aligned} x + 3y &= -2x \\ 4x + 2y &= -2y \end{aligned}$$

which consists of two identical equations

$$\begin{aligned} 3x + 3y &= 0 \\ 4x + 4y &= 0 \end{aligned}$$

Thus, $x + y = 0$. Let $y = t$; then $x = -t$, and any vector of the form

$$\mathbf{v} = \begin{bmatrix} x \\ y \end{bmatrix} = \begin{bmatrix} -t \\ t \end{bmatrix} = t \begin{bmatrix} -1 \\ 1 \end{bmatrix}$$

with $t \neq 0$ is an eigenvector corresponding to $\lambda = -2$. To check, we compute

$$A(\mathbf{v}) = \begin{bmatrix} 1 & 3 \\ 4 & 2 \end{bmatrix} \begin{bmatrix} -t \\ t \end{bmatrix} = \begin{bmatrix} -t + 3t \\ -4t + 2t \end{bmatrix} = \begin{bmatrix} 2t \\ -2t \end{bmatrix}$$

$$\lambda_2 \mathbf{v} = -2 \begin{bmatrix} -t \\ t \end{bmatrix} = \begin{bmatrix} 2t \\ -2t \end{bmatrix}$$

Thus, $A(\mathbf{v}) = -2\mathbf{v}$ for all vectors $\mathbf{v} = [-t \quad t]$, $t \neq 0$. ▲

A line through the origin is called *invariant* of a linear transformation A if A does not change it (i.e., does not change its direction). We visualize eigenvalues and eigenvectors by drawing *invariant lines* (or *invariant directions*) and indicating eigenvectors on them. See Figure 11.1.

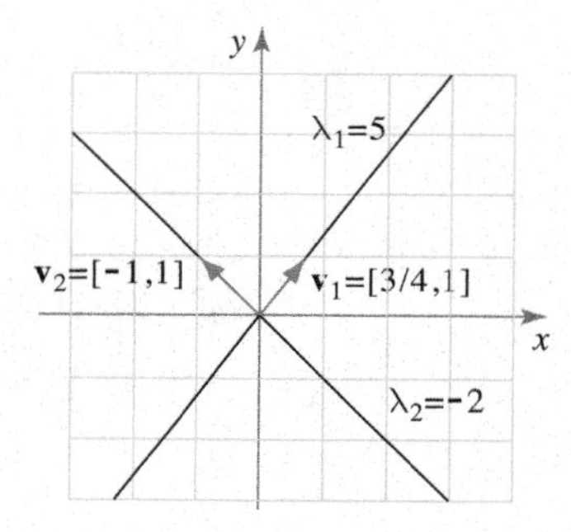

FIGURE 11.1
Invariant lines

In our case, the line

$$\begin{bmatrix} x \\ y \end{bmatrix} = t \begin{bmatrix} 3/4 \\ 1 \end{bmatrix}$$

of slope $1/(3/4) = 4/3$ and the line

$$\begin{bmatrix} x \\ y \end{bmatrix} = t \begin{bmatrix} -1 \\ 1 \end{bmatrix}$$

of slope $1/(-1) = -1$ are invariant.

Having investigated an example, we now describe the process of finding eigenvalues and eigenvectors for a general 2×2 matrix

$$A = \begin{bmatrix} a & b \\ c & d \end{bmatrix}$$

Recall that we are looking for a non-zero vector $\mathbf{v} = [x \quad y]$ and a real number λ such that $A\mathbf{v} = \lambda\mathbf{v}$. The matrix equation

$$A\begin{bmatrix} x \\ y \end{bmatrix} = \lambda \begin{bmatrix} x \\ y \end{bmatrix}$$

$$\begin{bmatrix} a & b \\ c & d \end{bmatrix}\begin{bmatrix} x \\ y \end{bmatrix} = \lambda \begin{bmatrix} x \\ y \end{bmatrix}$$

yields the system

$$\begin{aligned} ax + by &= \lambda x \\ cx + dy &= \lambda y \end{aligned}$$

Simplifying, we obtain a homogeneous linear system in two variables:

$$\begin{aligned} (a - \lambda)x + by &= 0 \\ cx + (d - \lambda)y &= 0 \end{aligned} \tag{11.5}$$

The matrix of the system

$$\begin{bmatrix} a - \lambda & b \\ c & d - \lambda \end{bmatrix}$$

can be written in the form

$$\begin{bmatrix} a - \lambda & b \\ c & d - \lambda \end{bmatrix} = \begin{bmatrix} a & b \\ c & d \end{bmatrix} + \begin{bmatrix} -\lambda & 0 \\ 0 & -\lambda \end{bmatrix} = A - \lambda I \tag{11.6}$$

where I is the 2×2 identity matrix.

We could have arrived at (11.6) without writing it all out in coordinates: start with

$$\begin{aligned} A \cdot \mathbf{v} &= \lambda\mathbf{v} \\ A \cdot \mathbf{v} - \lambda\mathbf{v} &= \mathbf{0} \end{aligned}$$

In order to factor out $\mathbf{v}$, we write $\mathbf{v} = I \cdot \mathbf{v}$, where I denotes the 2×2 identity matrix. Thus,

$$A \cdot \mathbf{v} - \lambda I \cdot \mathbf{v} = \mathbf{0}$$

$$(A - \lambda I) \cdot \mathbf{v} = \mathbf{0} \tag{11.7}$$

If we write (11.7) in coordinates, we obtain the system (11.5).

Theorem 10 in Section 9 tell us that the system in (11.5) or (11.7) has a non-trivial solution for $\mathbf{v} = [x \quad y]$ if and only if the determinant of the coefficient matrix (11.6) is zero:

$$\det(A - \lambda I) = 0 \tag{11.8}$$

Expanding the determinant, we obtain

$$\begin{vmatrix} a - \lambda & b \\ c & d - \lambda \end{vmatrix} = 0$$

$$(a - \lambda)(d - \lambda) - bc = 0$$

$$ad - d\lambda - a\lambda + \lambda^2 - bc = 0$$

$$\lambda^2 - (a + d)\lambda + (ad - bc) = 0 \tag{11.9}$$

Equation (11.9) is called the *characteristic equation of the matrix*

$$A = \begin{bmatrix} a & b \\ c & d \end{bmatrix}$$

(or the *characteristic equation of the transformation* represented by A). Note that the constant term $ad - bc$ in (11.9) is the determinant of A.

The sum of the diagonal terms in a square matrix is called the *trace* of A and is denoted by $\mathrm{tr}A$. In our case, $\mathrm{tr}A = a + d$, so we can rewrite the characteristic equation as

$$\lambda^2 - (\mathrm{tr}A)\lambda + \det A = 0 \tag{11.10}$$

The form (11.10) of the characteristic equation is quite convenient, as we can write it directly from the matrix A with a minimum of calculations (instead of repeating the steps that led to (11.9) from (11.8)).

In conclusion, the eigenvalues of a 2×2 matrix are the solutions of the quadratic equation (11.9) or (11.10). This means that a 2×2 matrix can have

(a) two distinct real eigenvalues,

(b) one (repeated) real eigenvalue, or

(c) two complex (conjugate) eigenvalues.

In this module, we discuss case (a) only. The remaining cases, as well as matrices of size larger than 2×2, are discussed in courses on linear algebra.

Example 11.3 Calculating Eigenvalues and Eigenvectors

Find the eigenvalues and the eigenvectors of the matrix

$$A = \begin{bmatrix} 4 & -3 \\ 0 & -3 \end{bmatrix}$$

▶ The characteristic equation of A is

$$\det(A - \lambda I) = \begin{vmatrix} 4-\lambda & -3 \\ 0 & -3-\lambda \end{vmatrix} = 0$$

i.e.,

$$(4-\lambda)(-3-\lambda) = 0$$

Thus, the eigenvalues of A are $\lambda_1 = 4$ and $\lambda_2 = -3$.

To find an eigenvector corresponding to $\lambda_1 = 4$, we need to solve the equation

$$A\mathbf{v} = 4\mathbf{v}$$

$$\begin{bmatrix} 4 & -3 \\ 0 & -3 \end{bmatrix} \begin{bmatrix} x \\ y \end{bmatrix} = 4 \begin{bmatrix} x \\ y \end{bmatrix}$$

for the coordinates x and y of the vector $\mathbf{v}$. The corresponding linear system

$$\begin{aligned} 4x - 3y &= 4x \\ -3y &= 4y \end{aligned}$$

simplifies to

$$\begin{aligned} 3y &= 0 \\ 7y &= 0 \end{aligned}$$

and thus $y = 0$. There are no conditions on x. Taking x to be a parameter t, we obtain

$$\mathbf{v} = \begin{bmatrix} t \\ 0 \end{bmatrix} = t \begin{bmatrix} 1 \\ 0 \end{bmatrix}$$

which gives all possible eigenvectors (assuming that $t \neq 0$) corresponding to the eigenvalue $\lambda_1 = 4$.

For $\lambda_2 = -3$, we compute

$$A\mathbf{v} = -3\mathbf{v}$$

$$\begin{bmatrix} 4 & -3 \\ 0 & -3 \end{bmatrix} \begin{bmatrix} x \\ y \end{bmatrix} = -3 \begin{bmatrix} x \\ y \end{bmatrix}$$

$$\begin{aligned} 4x - 3y &= -3x \\ -3y &= -3y \end{aligned}$$

This system reduces to a single equation

$$7x - 3y = 0$$

Let $y = t$; then $x = 3t/7$, and so every vector of the form

$$\mathbf{v} = \begin{bmatrix} 3t/7 \\ t \end{bmatrix} = t \begin{bmatrix} 3/7 \\ 1 \end{bmatrix}$$

is an eigenvector for $\lambda_2 = -3$.

Thus, the invariant lines (invariant directions) for A are the x-axis (corresponding to the eigenvalue $\lambda_1 = 4$) and the line through the origin of slope $7/3$ (corresponding to the eigenvalue $\lambda_2 = -3$); see Figure 11.2.

Under the transformation A, any vector lying on the invariant line $\lambda_1 = 4$ is scaled by a factor of 4. Any vector on the invariant line $\lambda_2 = -3$ is scaled by a factor of 3 and then reflected with respect to the origin. ▲

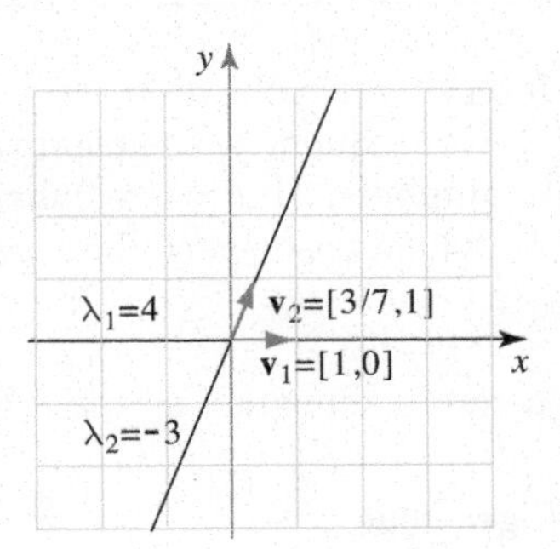

FIGURE 11.2

Invariant lines for the transformation in Example 11.3

Looking more closely at the previous example, we see that the eigenvalues of A are its diagonal entries. This is true in general for certain types of matrices. Assume that

$$A = \begin{bmatrix} a & b \\ c & d \end{bmatrix}$$

where either $b = 0$ or $c = 0$ or both $b = 0$ and $c = 0$. The eigenvalues of A are the zeros of the characteristic equation $\det(A - \lambda I) = 0$; thus

$$\begin{vmatrix} a - \lambda & b \\ c & d - \lambda \end{vmatrix} = 0$$

$$(a - \lambda)(d - \lambda) - bc = 0$$

$$(a - \lambda)(d - \lambda) = 0$$

since, by assumption, $bc = 0$. Thus, the eigenvalues are $\lambda_1 = a$ and $\lambda_2 = d$.

In conclusion, if a matrix

$$A = \begin{bmatrix} a & b \\ c & d \end{bmatrix}$$

is

(a) upper triangular ($c = 0$), or

(b) lower triangular ($b = 0$), or

(c) diagonal ($b = 0$ and $c = 0$),

then the eigenvalues of A are the diagonal entries a and d.

Example 11.4 Eigenvalues of the Rotation Matrix

Consider a rotation by an angle θ about the origin in the xy-plane, given by

$$A = \begin{bmatrix} \cos\theta & -\sin\theta \\ \sin\theta & \cos\theta \end{bmatrix}$$

(see Example 10.3 in Section 10).

We have already mentioned that if θ is not an integer multiple of π, then A does not have an invariant direction. Let's see what happens when we try to find its eigenvalues.

We start with the characteristic equation

$$\lambda^2 - (\mathrm{tr}A)\lambda + \det A = 0$$

The trace of A is $\mathrm{tr}A = 2\cos\theta$, and its determinant is $\det A = \cos^2\theta + \sin^2\theta = 1$. Thus

$$\lambda^2 - (2\cos\theta)\lambda + 1 = 0$$

Using the quadratic formula,

$$\lambda = \frac{2\cos\theta \pm \sqrt{4\cos^2\theta - 4}}{2} = \cos\theta \pm \sqrt{\cos^2\theta - 1}$$

Since θ is not an integer multiple of π, $\cos^2\theta < 1$ and thus the discriminant $\cos^2\theta - 1$ is negative. Consequently, the eigenvalues are not real—they are a pair of complex conjugate numbers. (If $\theta = k\pi$, where k is an integer, then the rotation is for 180° or 360°, in which case every direction is invariant.) ▲

A matrix can have a zero eigenvalue, as the following example shows.

Example 11.5 Matrix with a Zero Eigenvalue

Find the eigenvalues and eigenvectors of the matrix

$$A = \begin{bmatrix} 0 & 0 \\ 0 & 1 \end{bmatrix}$$

▶ The matrix A defines the transformation

$$A\begin{bmatrix} x \\ y \end{bmatrix} = \begin{bmatrix} 0 & 0 \\ 0 & 1 \end{bmatrix}\begin{bmatrix} x \\ y \end{bmatrix} = \begin{bmatrix} 0 \\ y \end{bmatrix}$$

A vector $[x \quad y]$ is transformed to the vector $[0 \quad y]$; i.e., the y coordinate is preserved, whereas the x coordinate changes to 0 (and so the image vector $[0 \quad y]$ lies on the y-axis). We conclude that A is the orthogonal projection onto the y-axis; see Figure 11.3.

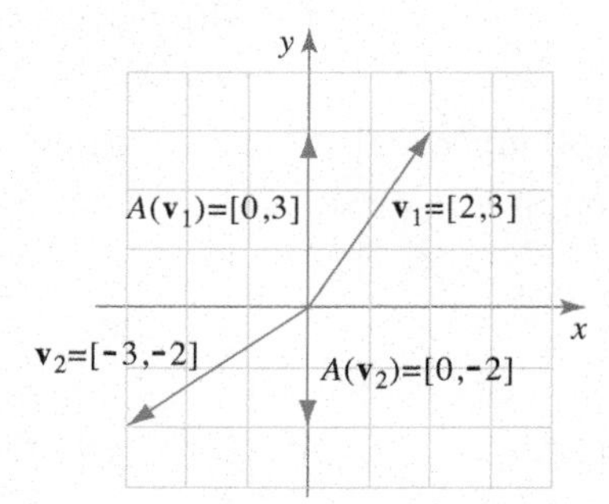

FIGURE 11.3
Orthogonal projection onto the y-axis

Any vector that is not vertical sees its direction changed when projected onto the y-axis. The vectors sitting on the y-axis remain on it, and therefore the y-axis is an invariant direction. Vectors parallel to the x-axis project to the origin (i.e., a zero vector on the y-axis).

Since A is a diagonal matrix, its eigenvalues are $\lambda_1 = 0$ and $\lambda_2 = 1$. The eigenvectors corresponding to $\lambda_1 = 0$ are the solutions of the system

$$A\begin{bmatrix} x \\ y \end{bmatrix} = 0\begin{bmatrix} x \\ y \end{bmatrix}$$

$$\begin{bmatrix} 0 & 0 \\ 0 & 1 \end{bmatrix}\begin{bmatrix} x \\ y \end{bmatrix} = \begin{bmatrix} 0 \\ 0 \end{bmatrix}$$

$$0 \cdot x + 0 \cdot y = 0$$
$$y = 0$$

Since x can be any real number, we use a parameter and write $x = t$. The eigenvectors

$$\mathbf{v} = \begin{bmatrix} t \\ 0 \end{bmatrix} = t \begin{bmatrix} 1 \\ 0 \end{bmatrix}$$

lie on the x-axis. Under A, all vectors on the x-axis project to the zero vector (the origin). Since the whole x-axis collapses to a point, we cannot say that the x-axis is an invariant *line* (this is an anomaly that occurs only when an eigenvalue is zero).

The eigenvectors corresponding to $\lambda_2 = 1$ are found in the same way:

$$A \begin{bmatrix} x \\ y \end{bmatrix} = 1 \begin{bmatrix} x \\ y \end{bmatrix}$$
$$\begin{bmatrix} 0 & 0 \\ 0 & 1 \end{bmatrix} \begin{bmatrix} x \\ y \end{bmatrix} = \begin{bmatrix} x \\ y \end{bmatrix}$$
$$0 = x$$
$$y = y$$

Thus, $x = 0$ and $y = t$, where t is a parameter. We conclude that all eigenvectors for $\lambda_2 = 1$ are of the form

$$\mathbf{v} = \begin{bmatrix} 0 \\ t \end{bmatrix} = t \begin{bmatrix} 0 \\ 1 \end{bmatrix}$$

where t is a non-zero real number. The invariant line is the y-axis, which agrees with our geometric reasoning. ▲

Instead of saying "all eigenvectors are of the form

$$\mathbf{v} = t \begin{bmatrix} 0 \\ 1 \end{bmatrix}$$

where t is a non-zero real number" we can say "an eigenvector is $\mathbf{v} = [0 \quad 1]$," since all other eigenvectors are parallel to it.

With this is mind, instead of using $[3/7 \quad 1]$ for an eigenvector (in Example 11.3), we could have used, for convenience, the vector $[3 \quad 7]$ or other multiples of $[3/7 \quad 1]$ that produce integer coordinates.

Eigenvalues, Eigenvectors, and Systems of Differential Equations

From calculus, we know that the solution of the differential equation

$$x'(t) = kx(t) \tag{11.11}$$

where k is a non-zero constant, is the exponential function $x(t) = Ce^{kt}$. The value of the constant C is determined from the initial condition

$$x(0) = Ce^{k0} = C$$

Thus, the solution of (11.11) is

$$x(t) = x(0)e^{kt} \tag{11.12}$$

Now consider a *system* of differential equations

$$\begin{aligned} x'(t) &= 2x(t) \\ y'(t) &= -3y(t) \end{aligned} \tag{11.13}$$

Since each equation involves only one function, we solve them separately, using (11.12):

$$\begin{aligned} x(t) &= x(0)e^{2t} \\ y(t) &= y(0)e^{-3t} \end{aligned} \tag{11.14}$$

Defining the vector

$$\mathbf{v}(t) = \begin{bmatrix} x(t) \\ y(t) \end{bmatrix}$$

we can write the system (11.13) in matrix form as

$$\begin{bmatrix} x'(t) \\ y'(t) \end{bmatrix} = \begin{bmatrix} 2 & 0 \\ 0 & -3 \end{bmatrix} \begin{bmatrix} x(t) \\ y(t) \end{bmatrix}$$

i.e., as $\mathbf{v}'(t) = A\mathbf{v}(t)$, where A is the matrix

$$A = \begin{bmatrix} 2 & 0 \\ 0 & -3 \end{bmatrix}$$

In vector notation, the solution (11.14) takes on the form

$$\begin{aligned} \mathbf{v}(t) = \begin{bmatrix} x(t) \\ y(t) \end{bmatrix} &= \begin{bmatrix} x(0)e^{2t} \\ y(0)e^{-3t} \end{bmatrix} \\ &= \begin{bmatrix} x(0)e^{2t} \\ 0 \end{bmatrix} + \begin{bmatrix} 0 \\ y(0)e^{-3t} \end{bmatrix} \\ &= x(0)e^{2t} \begin{bmatrix} 1 \\ 0 \end{bmatrix} + y(0)e^{-3t} \begin{bmatrix} 0 \\ 1 \end{bmatrix} \end{aligned} \tag{11.15}$$

Thus, the solution vector $\mathbf{v}(t)$ is a linear combination (a bit different, since the coefficients are not real numbers but functions of t) of the vectors $[1 \quad 0]$ and $[0 \quad 1]$.

Note that the coefficients in the exponents in (11.15) are the eigenvalues 2 and -3 of A. (Recall that the eigenvalues of a diagonal matrix are the diagonal entries.) In Exercise 19 we show that $[1 \quad 0]$ is an eigenvector for $\lambda_1 = 2$, and $[0 \quad 1]$ is an eigenvector for $\lambda_2 = -3$. Thus, each term in the solution (11.15) is of the form

$$\text{constant} \cdot e^{\text{eigenvalue} \cdot t} \cdot \text{eigenvector}$$

We have just discovered an important result, which we formulate as a theorem. Although we state it for 2×2 matrices, the theorem holds (under the same assumptions) for a general system of n linear differential equations with n functions.

Consider the system of linear differential equations

$$\mathbf{v}'(t) = A\mathbf{v}(t) \tag{11.16}$$

where

$$\mathbf{v}(t) = \begin{bmatrix} x(t) \\ y(t) \end{bmatrix}$$

and A is a 2×2 matrix.

Theorem 11 **Solution of a System of Linear Differential Equations**

Assume that A has two real, distinct eigenvalues λ_1 and λ_2, and let $\mathbf{v}_1$ and $\mathbf{v}_2$ be the corresponding eigenvectors. The general solution of the system of linear differential equations (11.16) is given by

$$\mathbf{v}(t) = C_1 e^{\lambda_1 t} \mathbf{v}_1 + C_2 e^{\lambda_2 t} \mathbf{v}_2$$

where C_1 and C_2 are constants.

Example 11.6 Solving a System of Differential Equations

Find the general solution of the system

$$x'(t) = 3x(t) + 8y(t)$$
$$y'(t) = x(t) - 4y(t)$$

▶ We write the given system as $\mathbf{v}'(t) = A\mathbf{v}(t)$, where $\mathbf{v}(t) = [x(t) \quad y(t)]$ and

$$A = \begin{bmatrix} 3 & 8 \\ 1 & -4 \end{bmatrix}.$$

is the matrix of the system.

Using (11.10), we write the characteristic equation as

$$\lambda^2 + \lambda - 20 = 0$$
$$(\lambda - 4)(\lambda + 5) = 0$$

We conclude that A has two distinct real eigenvalues, $\lambda_1 = 4$ and $\lambda_2 = -5$. To find an eigenvector corresponding to $\lambda_1 = 4$, we have to solve the system

$$\begin{bmatrix} 3 & 8 \\ 1 & -4 \end{bmatrix} \begin{bmatrix} x \\ y \end{bmatrix} = 4 \begin{bmatrix} x \\ y \end{bmatrix}$$

$$3x + 8y = 4x$$
$$x - 4y = 4y$$

$$-x + 8y = 0$$
$$x - 8y = 0$$

The two equations are identical. We need one eigenvector (and not a general form involving a parameter), so let $y = 1$. Then $x = 8$, and $\mathbf{v}_1 = [8 \quad 1]$ is an eigenvector for $\lambda_1 = 4$.

Likewise, for $\lambda_2 = -5$ we obtain

$$\begin{bmatrix} 3 & 8 \\ 1 & -4 \end{bmatrix} \begin{bmatrix} x \\ y \end{bmatrix} = -5 \begin{bmatrix} x \\ y \end{bmatrix}$$

$$3x + 8y = -5x$$
$$x - 4y = -5y$$

$$8x + 8y = 0$$
$$x + y = 0$$

Again, the system collapses to one equation, $x + y = 0$. Let $x = 1$; then $y = -1$; the vector $\mathbf{v}_2 = [1 \quad -1]$ is an eigenvector for $\lambda_2 = -5$.

We are done: the solution to the given system is

$$\mathbf{v}(t) = C_1 e^{4t} \begin{bmatrix} 8 \\ 1 \end{bmatrix} + C_2 e^{-5t} \begin{bmatrix} 1 \\ -1 \end{bmatrix}$$

or, writing out the coordinates,

$$x(t) = 8C_1 e^{4t} + C_2 e^{-5t}$$
$$y(t) = C_1 e^{4t} - C_2 e^{-5t}$$

The constants C_1 and C_2 need to be determined from the initial conditions. ▲

The system we just solved is an example of a linear system describing the dynamics of competing species, which was introduced at the beginning of Section 5.

Summary A non-zero vector $\mathbf{v}$ is an **eigenvector** of a matrix A (or of a linear transformation A) if $A\mathbf{v} = \lambda\mathbf{v}$ (or $A(\mathbf{v}) = \lambda\mathbf{v}$), that is, if A transforms a vector to its multiple.

The number λ is called an **eigenvalue.** To find eigenvalues we solve the **characteristic equation,** which, in the case of 2×2 matrices, is a quadratic equation. The corresponding eigenvectors are the solutions of the underdetermined system $A(\mathbf{v}) = \lambda\mathbf{v}$. The eigenvalues and eigenvectors help us solve systems of linear differential equations. An application to population dynamics is discussed in the next section.

11 Exercises

1. Give an example of a 2×2 matrix whose eigenvalues are 3 and 7.

2. Find a 2×2 matrix whose eigenvalues are 1 and -5, and which has at least three non-zero entries.

3. Find a 2×2 matrix with non-zero entries whose eigenvalues are 1 and -3.

4. Find a 2×2 matrix with non-zero entries whose eigenvalues are 4 and 2.

5–8 ▪ Based on a geometric interpretation of the linear transformation defined by each matrix A, find the eigenvectors and the corresponding eigenvalues of A.

5. $A = \begin{bmatrix} 2 & 0 \\ 0 & 2 \end{bmatrix}$

6. $A = \begin{bmatrix} 1 & 0 \\ 0 & 0 \end{bmatrix}$

7. $A = \begin{bmatrix} 1 & 0 \\ 0 & -1 \end{bmatrix}$

8. $A = \begin{bmatrix} 2 & 0 \\ 0 & 3 \end{bmatrix}$

9. Let

$$A = \begin{bmatrix} 0 & 2 \\ 3 & 1 \end{bmatrix}, \mathbf{v}_1 = \begin{bmatrix} 4 \\ 4 \end{bmatrix}, \mathbf{v}_2 = \begin{bmatrix} -2 \\ 2 \end{bmatrix}, \mathbf{v}_3 = \begin{bmatrix} 1 \\ 2 \end{bmatrix}$$

Determine whether or not each of the vectors $\mathbf{v}_1$, $\mathbf{v}_2$, and $\mathbf{v}_3$ is an eigenvector of the matrix A. For each eigenvector, find the corresponding eigenvalue.

10. Let

$$A = \begin{bmatrix} 1 & 2 \\ 3 & 2 \end{bmatrix}, \mathbf{v}_1 = \begin{bmatrix} -2 \\ -3 \end{bmatrix}, \mathbf{v}_2 = \begin{bmatrix} -1 \\ 1 \end{bmatrix}, \mathbf{v}_3 = \begin{bmatrix} 0 \\ 3 \end{bmatrix}$$

Determine whether or not each of the vectors $\mathbf{v}_1$, $\mathbf{v}_2$, and $\mathbf{v}_3$ is an eigenvector of the matrix A. For each eigenvector, find the corresponding eigenvalue.

11–18 ▪ Find the eigenvalues and the corresponding eigenvectors for each matrix.

11. $\begin{bmatrix} 7 & 0 \\ 0 & -5 \end{bmatrix}$

12. $\begin{bmatrix} -4 & 0 \\ 0 & 1 \end{bmatrix}$

13. $\begin{bmatrix} 1 & 2 \\ 2 & 4 \end{bmatrix}$

14. $\begin{bmatrix} 5 & 0 \\ 1 & -2 \end{bmatrix}$

15. $\begin{bmatrix} 11 & -2 \\ 12 & 1 \end{bmatrix}$

16. $\begin{bmatrix} 4 & -3 \\ -3 & 4 \end{bmatrix}$

17. $\begin{bmatrix} 1.5 & 3.5 \\ 3.5 & 1.5 \end{bmatrix}$

18. $\begin{bmatrix} -0.25 & 5.25 \\ 1.75 & 3.25 \end{bmatrix}$

19. The eigenvalues of the matrix

$$\begin{bmatrix} 2 & 0 \\ 0 & -3 \end{bmatrix}$$

are 2 and -3. Show that the corresponding eigenvectors are $\begin{bmatrix} 1 & 0 \end{bmatrix}$ and $\begin{bmatrix} 0 & 1 \end{bmatrix}$.

20. If 4 is an eigenvalue of the matrix A, is it true that 8 is an eigenvalue of the matrix $2A$?

21. If 4 is an eigenvalue of the matrix A, is it true that -4 is an eigenvalue of the matrix $-A$?

22. Sketch the invariant directions of the matrix

$$A = \begin{bmatrix} 2 & -4 \\ 2 & -4 \end{bmatrix}$$

and identify them by the corresponding eigenvalue.

23. Sketch the invariant directions of the matrix

$$A = \begin{bmatrix} -9 & 6 \\ -12 & 9 \end{bmatrix}$$

and identify them by the corresponding eigenvalue.

24. Assume that λ_1 and λ_2 are the eigenvalues of a 2×2 matrix A. Show that $\lambda_1 + \lambda_2 = -\text{tr}A$ and $\lambda_1 \lambda_2 = \det A$.

25. Assume that $\mathbf{v}$ is an eigenvector of both matrices A and B. Show that $\mathbf{v}$ is also an eigenvector of $A + B$, and find the corresponding eigenvalue.

26. Assume that $\mathbf{v}$ is an eigenvector of both square matrices A and B. Show that $\mathbf{v}$ is also an eigenvector of their product AB, and find the corresponding eigenvalue.

27. Assume that λ is an eigenvalue of a matrix A, with the corresponding eigenvector $\mathbf{v}$. Show that λ^2 is an eigenvalue for A^2, and find the corresponding eigenvector.

28. Assume that A is an invertible 2×2 matrix. Let λ be an eigenvalue of A, with the corresponding eigenvector $\mathbf{v}$. Prove that $\mathbf{v}$ is an eigenvector of the inverse matrix A^{-1}, and find the corresponding eigenvalue.

29–32 ▪ Find the general solution of each system of differential equations.

29. $x'(t) = x(t) + 2y(t)$
$y'(t) = 2x(t) + 4y(t)$

30. $x'(t) = x(t) + 2y(t)$
$y'(t) = 3x(t) + 2y(t)$

31. $x'(t) = 3x(t) + 8y(t)$
$y'(t) = -y(t)$

32. $x'(t) = 4x(t) + 6y(t)$
$y'(t) = y(t)$

12 The Leslie Model: Age-Structured Population Dynamics

Using the linear algebra that we have learned, in particular **matrix multiplication, eigenvalues,** and **eigenvectors,** we develop a model that will help us understand the dynamics of **age distribution** within a population.

Introduction

The simplest model of a one-species population dynamics is given by

$$p(t+1) = rp(t) \tag{12.1}$$

where $p(t)$ is the population size at time t. The start of the experiment (observation) is labelled by $t = 0$, and $p(0)$ represents the initial population (which is usually known). The constant r is the *per capita production rate;* it is the number of individuals at time $t+1$ per individual at time t. Or, writing (12.1) as $r = p(t+1)/p(t)$, we interpret r as the *relative change* in population between the times t and $t+1$.

If $r > 1$, the population increases, and if $r < 1$ the population decreases. When $r = 1$ there is no change in size: every member that dies is immediately replaced by a new member.

From $p(t+1) = rp(t)$ we compute

$$\begin{aligned} p(1) &= rp(0) \\ p(2) &= rp(1) = r^2p(0) \\ p(3) &= rp(2) = r^3p(0) \end{aligned}$$

and, in general, $p(t) = r^tp(0)$. Thus, the model (12.1) predicts that the population will grow or decay exponentially (or remain unchanged).

By assuming that the per capita rate r is constant in (12.1), one important aspect is missed: quite often, the reproduction rate depends on the age of an individual. At different stages in life, the reproduction rates could be quite different.

In 1945, Patrick Leslie built a model that accounts for reproduction based on age. [See Leslie, P. (1945). On the use of matrices in certain population mathematics. *Biometrika,* 33 (3), 183–212.] To explain how the model works, we start with an example.

Consider a population of bacteria that multiply once a month. We divide the bacteria into four age groups (also called age classes): baby bacteria, juvenile bacteria, adult bacteria, and senior bacteria. We assume that each age group spans a month: for instance, a juvenile bacterium will become an adult bacterium some time during the following month; as well, an adult bacterium will become a senior bacterium the following month. The lifespan of the bacteria is assumed to be four months; thus, all bacteria of (currently) senior age will die the following month.

By t we denote the time measured in months; in this way, the integer values of t count the number of generations. To keep track of the number of individuals in each age group, we define the vector

$$P(t) = \begin{bmatrix} \text{number of baby bacteria at time } t \\ \text{number of juvenile bacteria at time } t \\ \text{number of adult bacteria at time } t \\ \text{number of senior bacteria at time } t \end{bmatrix} = \begin{bmatrix} P_1(t) \\ P_2(t) \\ P_3(t) \\ P_4(t) \end{bmatrix}$$

The total population at time t is $P_1(t) + P_2(t) + P_3(t) + P_4(t)$.

To define the model, we need to know *demographic data:*

(a) How do bacteria age?

(b) How do they reproduce?

(Getting good, reliable demographic information is usually the toughest part of population research.)

For (a), we need the *survival probabilities,* i.e., the chance that a bacterium living today will be alive a month from now. We assume that 60% of baby bacteria survive the first month and become juvenile bacteria, 40% of juvenile bacteria live another month and become adult bacteria, and 25% of adult bacteria live a month longer and become senior bacteria.

For (b), we need the *birth parameters;* i.e., we need to know which bacteria are capable of reproducing and how many offspring they produce. Assume that neither the baby nor the senior bacteria are capable of reproducing. The per capita reproduction rate is 1.3 for the juveniles and 3 for the adults.

Finally, assume that the demographic data do not change from generation to generation (in our case, from month to month).

Example 12.1 Sample Calculations with the Model

Assume that, currently, there are 800 baby bacteria, 500 juvenile bacteria, 200 adult bacteria, and 60 senior bacteria. How many bacteria will there be in each age group in the following month?

▶ Of 800 baby bacteria, 60%, or 480, will live to become juvenile. Of 500 juvenile bacteria, 40%, or 200, will become adults. Of 200 adult bacteria, 25%, or 50, will become seniors. The 60 senior bacteria will die.

How many baby bacteria will there be? Each of 500 juveniles produces (on average) 1.5 bacteria, which is 750 baby bacteria; each of 200 adults produces on average 3 bacteria, which contributes 600 baby bacteria to the pool. So there will be a total of 1350 baby bacteria next month.

What happens the following month? After ten months? In the long term?

We will be able to answer these questions, but first we need to organize our calculations. We put the initial data into the vector

$$P(0) = \begin{bmatrix} 800 \\ 500 \\ 200 \\ 60 \end{bmatrix}$$

We calculated that

$$P(1) = \begin{bmatrix} 1350 \\ 480 \\ 200 \\ 50 \end{bmatrix} \tag{12.2}$$

We know that matrices transform vectors, so our next step is to find a matrix that transforms the vector $P(0)$ into the vector $P(1)$. Then we will use the same matrix to obtain $P(2)$ from $P(1)$, $P(3)$ from $P(2)$, and so on.

Given that

$$P(t) = \begin{bmatrix} P_1(t) \\ P_2(t) \\ P_3(t) \\ P_4(t) \end{bmatrix}$$

how do we find the age distribution $P(t+1)$?

It is given that 60% of baby bacteria (at time t) will become juvenile (at time $t+1$); thus $P_2(t+1) = 0.6P_1(t)$. In the same way, looking at the survival probabilities, we obtain the following equations:

$$\begin{aligned} P_2(t+1) &= 0.6P_1(t) \\ P_3(t+1) &= 0.4P_2(t) \\ P_4(t+1) &= 0.25P_3(t) \end{aligned} \tag{12.3}$$

The birth parameters information implies that

$$P_1(t+1) = 1.5P_2(t) + 3P_3(t) \tag{12.4}$$

We write (12.3) and (12.4) in matrix form as

$$\begin{bmatrix} P_1(t+1) \\ P_2(t+1) \\ P_3(t+1) \\ P_4(t+1) \end{bmatrix} = \begin{bmatrix} 0 & 1.5 & 3 & 0 \\ 0.6 & 0 & 0 & 0 \\ 0 & 0.4 & 0 & 0 \\ 0 & 0 & 0.25 & 0 \end{bmatrix} \begin{bmatrix} P_1(t) \\ P_2(t) \\ P_3(t) \\ P_4(t) \end{bmatrix} \tag{12.5}$$

The 4×4 matrix in (12.5) is called the *Leslie matrix* and is denoted by L. It fully describes the dynamics of the age-distributed population change. Let's see how it works.

The initial age distribution is given by the matrix

$$P(0) = \begin{bmatrix} 800 \\ 500 \\ 200 \\ 60 \end{bmatrix}$$

From (12.5) we compute the age distribution one month later:

$$\begin{bmatrix} P_1(1) \\ P_2(1) \\ P_3(1) \\ P_4(1) \end{bmatrix} = \begin{bmatrix} 0 & 1.5 & 3 & 0 \\ 0.6 & 0 & 0 & 0 \\ 0 & 0.4 & 0 & 0 \\ 0 & 0 & 0.25 & 0 \end{bmatrix} \begin{bmatrix} 800 \\ 500 \\ 200 \\ 60 \end{bmatrix} = \begin{bmatrix} (1.5)(500) + (3)(200) \\ (0.6)(800) \\ (0.4)(500) \\ (0.25)(200) \end{bmatrix} = \begin{bmatrix} 1350 \\ 480 \\ 200 \\ 50 \end{bmatrix}$$

which agrees with our earlier calculations and with (12.2). We continue in the same way; to find the age distribution at time $t+1$ we apply the Leslie matrix to the age distribution at the t.

The age distribution after two months is

$$\begin{bmatrix} P_1(2) \\ P_2(2) \\ P_3(2) \\ P_4(2) \end{bmatrix} = \begin{bmatrix} 0 & 1.5 & 3 & 0 \\ 0.6 & 0 & 0 & 0 \\ 0 & 0.4 & 0 & 0 \\ 0 & 0 & 0.25 & 0 \end{bmatrix} \begin{bmatrix} 1350 \\ 480 \\ 200 \\ 50 \end{bmatrix} = \begin{bmatrix} 1320 \\ 810 \\ 192 \\ 50 \end{bmatrix}$$

Continuing, we obtain

$$P(3) = \begin{bmatrix} 0 & 1.5 & 3 & 0 \\ 0.6 & 0 & 0 & 0 \\ 0 & 0.4 & 0 & 0 \\ 0 & 0 & 0.25 & 0 \end{bmatrix} \begin{bmatrix} 1320 \\ 810 \\ 192 \\ 50 \end{bmatrix} = \begin{bmatrix} 1791 \\ 792 \\ 324 \\ 48 \end{bmatrix},$$

and

$$P(4) = \begin{bmatrix} 2160 \\ 1075 \\ 317 \\ 81 \end{bmatrix}, \quad P(5) = \begin{bmatrix} 2564 \\ 1296 \\ 430 \\ 79 \end{bmatrix},$$

and so on (some entries have been rounded off). Using a computer, we calculated the age distribution for the first ten months; see Figure 12.1. As we can see, the size of each age group increases; thus, of course, the total population increases as well.

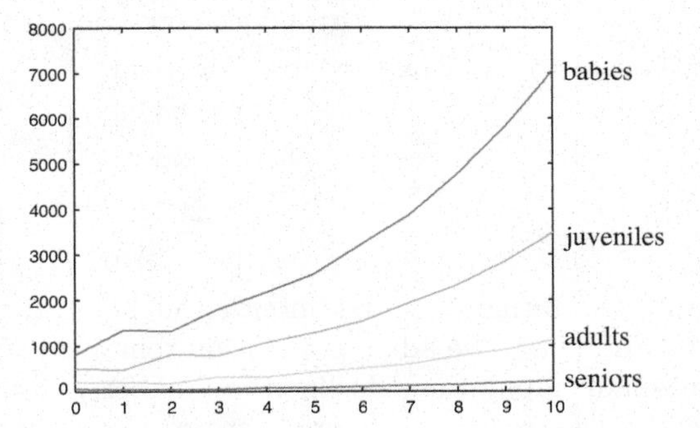

FIGURE 12.1

Age distribution of the bacteria population for the first ten months

To understand the implications of the model a bit better, instead of the actual (absolute) age group sizes we calculate the *relative age group sizes*

$$\begin{bmatrix} R_1(t) \\ R_2(t) \\ R_3(t) \\ R_4(t) \end{bmatrix} = \frac{1}{P_1(t) + P_2(t) + P_3(t) + P_4(t)} \begin{bmatrix} P_1(t) \\ P_2(t) \\ P_3(t) \\ P_4(t) \end{bmatrix}$$

For instance,

$$\begin{bmatrix} R_1(0) \\ R_2(0) \\ R_3(0) \\ R_4(0) \end{bmatrix} = \frac{1}{800 + 500 + 200 + 60} \begin{bmatrix} 800 \\ 500 \\ 200 \\ 60 \end{bmatrix} = \begin{bmatrix} 0.513 \\ 0.321 \\ 0.128 \\ 0.038 \end{bmatrix}$$

tells us that the initial population consists of 51.3% baby bacteria, 32.1% juvenile bacteria, 12.8% adult bacteria, and 3.8% senior bacteria. After the first month,

$$\begin{bmatrix} R_1(1) \\ R_2(1) \\ R_3(1) \\ R_4(1) \end{bmatrix} = \frac{1}{1350 + 480 + 200 + 50} \begin{bmatrix} 1350 \\ 480 \\ 200 \\ 50 \end{bmatrix} = \begin{bmatrix} 0.649 \\ 0.231 \\ 0.096 \\ 0.024 \end{bmatrix}$$

Figure 12.2 shows the relative age distribution for the four age groups. While the absolute sizes increase, the relative sizes seem to be approaching constant values.

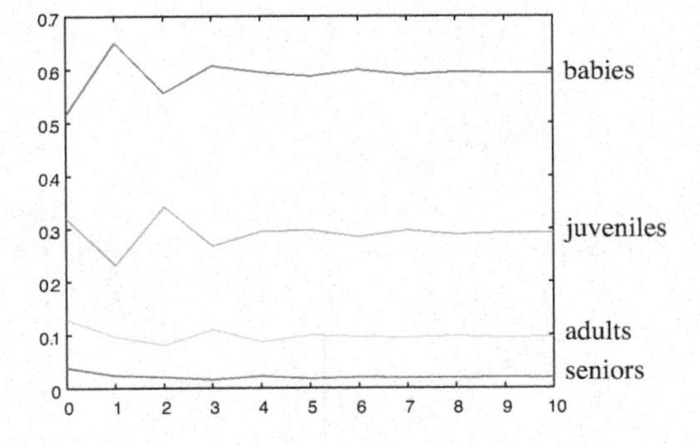

FIGURE 12.2

Relative age distribution of the bacteria for the first ten months

It can be shown that the relative populations $R_1(t), R_2(t), R_3(t)$, and $R_4(t)$ approach (in the limit, as $t \to \infty$) the *steady state*

$$R = \begin{bmatrix} 0.593 \\ 0.292 \\ 0.096 \\ 0.019 \end{bmatrix}$$

which represents the *stable age distribution.*

To quantify the changes within an individual age group, we look at the ratios

$$S_i(t) = \frac{\text{group size in the present generation (time } t)}{\text{group size one month ago (time } t-1)}$$

for each of the four age groups. For instance,

$$S(1) = \begin{bmatrix} 1350/800 \\ 480/500 \\ 200/200 \\ 50/60 \end{bmatrix} = \begin{bmatrix} 1.688 \\ 0.960 \\ 1 \\ 0.833 \end{bmatrix}$$

says that during the first month the baby bacteria population increased by a factor of 1.688, the adult population remained constant, whereas the juvenile and the senior populations declined in numbers. Figure 12.3 suggests that in the long term, all four ratios converge toward the same number. Using a computer, we obtain

$$S(10) = \begin{bmatrix} 1.216 \\ 1.220 \\ 1.231 \\ 1.194 \end{bmatrix}$$

It can be proven that all four numbers approach 1.221 as $t \to \infty$.

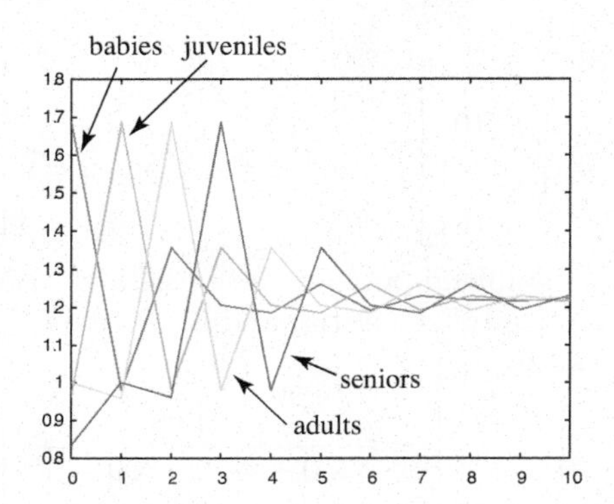

FIGURE 12.3

Ratio of the present to the previous month size for each age group

Using eigenvalues and eigenvectors a bit later in this section, we will be able to explain the behaviour of the age groups and obtain good insight into the phenomena that we observed in this example.

The Leslie Matrix

In general, instead of considering four age groups, we consider n age groups of equal timespan. From the given demographic data, we construct the $n \times n$ *Leslie matrix*

$$L = \begin{bmatrix} b_1 & b_2 & b_3 & \cdots & b_{n-1} & b_n \\ p_1 & 0 & 0 & \cdots & 0 & 0 \\ 0 & p_2 & 0 & \cdots & 0 & 0 \\ 0 & 0 & p_3 & \cdots & 0 & 0 \\ \vdots & \vdots & \vdots & \ddots & \vdots & \vdots \\ 0 & 0 & 0 & \cdots & p_{n-1} & 0 \end{bmatrix}$$

In his paper (mentioned in the introduction to this section), Leslie used $n = 21$ age groups (spanning one month each) of an imaginary population of 100 thousand brown rats. The numbers $b_1, b_2, \ldots, b_n$ are the *birth parameters* (which determine the number of offspring in the next time interval), and $p_1, p_2, \ldots, p_{n-1}$ are the *probabilities of survival* (p_i is the probability that an individual survives in moving from age group i to age group $i+1$). The age-distributed population at time t is given by

$$P(t) = \begin{bmatrix} P_1(t) \\ P_2(t) \\ \dots \\ P_n(t) \end{bmatrix}$$

The time unit is equal to the common timespan of the age groups. The matrix product

$$P(t+1) = LP(t) \tag{12.6}$$

gives the age distribution at time $t+1$.

Leslie's model (12.6) has been used in population biology to obtain, and to predict, important demographic characteristics of the way human, plant, and animal populations (as well as cells, bacteria, and other organisms) develop and age.

Eigenvalues and Eigenvectors of the Leslie Matrix

In Example 12.1 we noticed that the sizes of age groups show a certain type of stability (see Figures 12.2 and 12.3). In order to investigate these phenomena further, we need eigenvalues and eigenvectors.

To keep calculations to a minimum, we use a 2×2 Leslie matrix. Although of small size, the matrix will, nevertheless, reveal essential features of age-distributed population changes.

Take the Leslie matrix

$$L = \begin{bmatrix} 1.05 & 0.9 \\ 0.2 & 0 \end{bmatrix}$$

and assume that the initial population is

$$P(0) = \begin{bmatrix} 75 \\ 35 \end{bmatrix}$$

Looking at L, we see that the whole population is broken down into two age groups (assumed to be of equal timespan), which we name "minor" and "adult." A minor survives to become an adult with a 20% chance. A minor produces, on average, 1.05 offspring, whereas the average number of offspring per adult is 0.9.

Using L, we compute the age distribution in successive generations:

$$\begin{aligned} P(1) &= LP(0) = \begin{bmatrix} 1.05 & 0.9 \\ 0.2 & 0 \end{bmatrix} \begin{bmatrix} 75 \\ 35 \end{bmatrix} = \begin{bmatrix} 110.25 \\ 15 \end{bmatrix} \\ P(2) &= LP(1) = \begin{bmatrix} 129.3 \\ 22.0 \end{bmatrix} \\ P(3) &= LP(2) = \begin{bmatrix} 155.6 \\ 25.9 \end{bmatrix} \\ P(4) &= LP(3) = \begin{bmatrix} 186.7 \\ 31.1 \end{bmatrix} \end{aligned} \tag{12.7}$$

and so on. (Of course, to interpret the numbers we need to round off to the nearest integer.) Both age groups increase in size; see Figure 12.4a. In terms of relative numbers (size of each age group as a percent of the total population), we observe the same phenomenon as in Example 12.1: the two age groups stabilize at

$$R = \begin{bmatrix} 0.857 \\ 0.143 \end{bmatrix}$$

In other words, the stable age distribution consists of 85.7% minors and 14.3% adults (Figure 12.4b).

The relative growth rates within each age group (the number of minors (adults) at time t divided by the number of minors (adults) at time $t-1$) converge toward the same number (around 1.2; see Figure 12.4c).

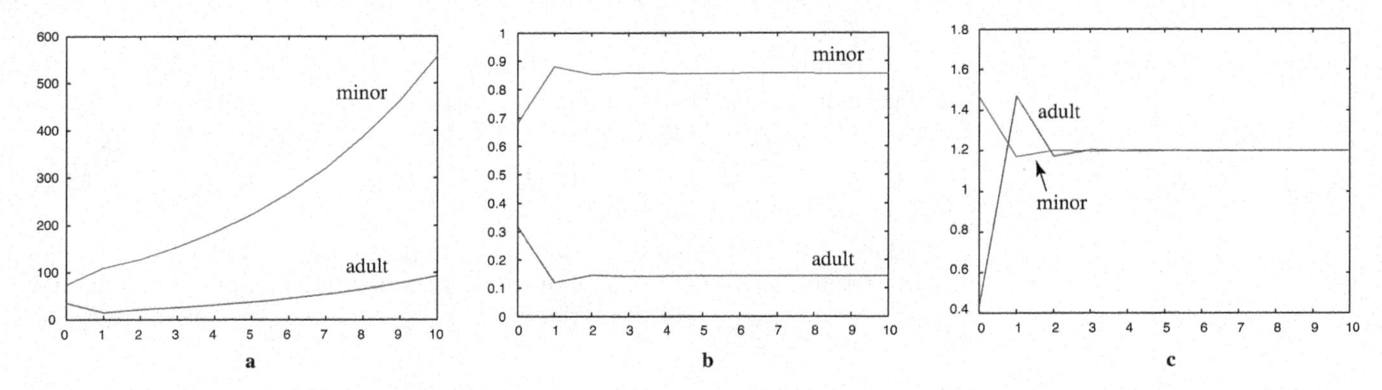

FIGURE 12.4
Age distribution in terms of absolute (a) and relative (b and c) population numbers

Using linear algebra, we now explain the behaviour of a population whose dynamics is defined by the Leslie model.

Example 12.2 Eigenvalues and Eigenvectors of a Leslie Matrix

Find the eigenvalues and eigenvectors of the Leslie matrix

$$L = \begin{bmatrix} 1.05 & 0.9 \\ 0.2 & 0 \end{bmatrix}$$

▶ Since $\text{tr}L = 1.05$ and $\det L = -0.18$, the characteristic polynomial of L is

$$\lambda^2 - 1.05\lambda - 0.18 = 0$$

(see (11.10) in Section 11). Factoring gives

$$(\lambda - 1.2)(\lambda + 0.15) = 0$$

so we obtain the eigenvalues $\lambda_1 = 1.2$ and $\lambda_2 = -0.15$. Note that 1.2 is equal to the limiting value of the relative growth rates within each age group (Figure 12.4c; maybe it's not a coincidence!).

To find an eigenvector corresponding to $\lambda_1 = 1.2$, we solve the system

$$\begin{bmatrix} 1.05 & 0.9 \\ 0.2 & 0 \end{bmatrix} \begin{bmatrix} x \\ y \end{bmatrix} = 1.2 \begin{bmatrix} x \\ y \end{bmatrix}$$

$$\begin{aligned} 1.05x + 0.9y &= 1.2x \\ 0.2x &= 1.2y \end{aligned}$$

$$\begin{aligned} 0.15x &= 0.9y \\ 0.2x &= 1.2y \end{aligned}$$

Both equations reduce to $x = 6y$. Pick $y = 1$; then $x = 6$ and

$$\begin{bmatrix} 6 \\ 1 \end{bmatrix}$$

is an eigenvector for the eigenvalue $\lambda_1 = 1.2$.

When $\lambda_2 = -0.15$, we obtain

$$\begin{bmatrix} 1.05 & 0.9 \\ 0.2 & 0 \end{bmatrix} \begin{bmatrix} x \\ y \end{bmatrix} = -0.15 \begin{bmatrix} x \\ y \end{bmatrix}$$

$$\begin{aligned} 1.05x + 0.9y &= -0.15x \\ 0.2x &= -0.15y \end{aligned}$$

$$\begin{aligned} 1.2x + 0.9y &= 0 \\ 0.2x + 0.15y &= 0 \end{aligned}$$

Both equations reduce to $4x + 3y = 0$. Take $y = 4$; then $x = -3$, and

$$\begin{bmatrix} -3 \\ 4 \end{bmatrix}$$

is an eigenvector corresponding to the eigenvalue $\lambda_2 = -0.15$. ▲

To proceed, we need a technical result.

Example 12.3 Iterated Maps

Assume that L is a 2×2 matrix with eigenvalues λ_1 and λ_2 and corresponding eigenvectors $\mathbf{v}_1$ and $\mathbf{v}_2$ (so that $L\mathbf{v}_1 = \lambda_1 \mathbf{v}_1$ and $L\mathbf{v}_2 = \lambda_2 \mathbf{v}_2$). Let u be a linear combination of $\mathbf{v}_1$ and $\mathbf{v}_2$; i.e.,

$$\mathbf{u} = \alpha_1 \mathbf{v}_1 + \alpha_2 \mathbf{v}_2 \tag{12.8}$$

where $\alpha_1, \alpha_2 \in \mathbb{R}$. Express the vectors $L\mathbf{u}$, $L^2\mathbf{u} = L \cdot (L\mathbf{u})$, $L^3\mathbf{u} = L \cdot (L^2\mathbf{u}), \ldots$ as a linear combination of the eigenvectors $\mathbf{v}_1$ and $\mathbf{v}_2$.

▶ Multiplying both sides of (12.8) from the left by L, we obtain

$$\begin{aligned} L\mathbf{u} &= L(\alpha_1 \mathbf{v}_1 + \alpha_2 \mathbf{v}_2) \\ &= L(\alpha_1 \mathbf{v}_1) + L(\alpha_2 \mathbf{v}_2) \\ &= \alpha_1 (L\mathbf{v}_1) + \alpha_2 (L\mathbf{v}_2) \end{aligned}$$

Using the fact that $\mathbf{v}_1$ and $\mathbf{v}_2$ are eigenvectors, we write:

$$\begin{aligned} L\mathbf{u} &= \alpha_1 (\lambda_1 \mathbf{v}_1) + \alpha_2 (\lambda_2 \mathbf{v}_2) \\ &= \alpha_1 \lambda_1 \mathbf{v}_1 + \alpha_2 \lambda_2 \mathbf{v}_2 \end{aligned} \tag{12.9}$$

The last line shows that $L\mathbf{u}$ is a linear combination of $\mathbf{v}_1$ and $\mathbf{v}_2$.

Multiplying (12.9) by L and proceeding as above, we obtain

$$\begin{aligned} L \cdot (L\mathbf{u}) &= L(\alpha_1 \lambda_1 \mathbf{v}_1 + \alpha_2 \lambda_2 \mathbf{v}_2) \\ &= L(\alpha_1 \lambda_1 \mathbf{v}_1) + L(\alpha_2 \lambda_2 \mathbf{v}_2) \\ &= \alpha_1 \lambda_1 (L\mathbf{v}_1) + \alpha_2 \lambda_2 (L\mathbf{v}_2) \\ &= \alpha_1 \lambda_1 (\lambda_1 \mathbf{v}_1) + \alpha_2 \lambda_2 (\lambda_2 \mathbf{v}_2) \\ &= \alpha_1 \lambda_1^2 \mathbf{v}_1 + \alpha_2 \lambda_2^2 \mathbf{v}_2 \end{aligned}$$

i.e.,

$$L^2\mathbf{u} = \alpha_1 \lambda_1^2 \mathbf{v}_1 + \alpha_2 \lambda_2^2 \mathbf{v}_2 \tag{12.10}$$

In the same way we show that

$$L^3\mathbf{u} = L(L^2\mathbf{u}) = \alpha_1 \lambda_1^3 \mathbf{v}_1 + \alpha_2 \lambda_2^3 \mathbf{v}_2$$

and, in general,

$$L^n\mathbf{u} = \alpha_1 \lambda_1^n \mathbf{v}_1 + \alpha_2 \lambda_2^n \mathbf{v}_2 \tag{12.11}$$

where $n = 1, 2, 3 \ldots$. ▲

What is the point of Example 12.3?

Given the Leslie matrix

$$L = \begin{bmatrix} 1.05 & 0.9 \\ 0.2 & 0 \end{bmatrix}$$

and the initial age distribution

$$P(0) = \begin{bmatrix} 75 \\ 35 \end{bmatrix}$$

we compute the age distribution at time $t = 1$ by matrix multiplication:

$$P(1) = LP(0) = \begin{bmatrix} 1.05 & 0.9 \\ 0.2 & 0 \end{bmatrix} \begin{bmatrix} 75 \\ 35 \end{bmatrix} = \begin{bmatrix} 110.25 \\ 15 \end{bmatrix}$$

To compute $P(2)$, we multiply L by $P(1)$; to find $P(3)$ we multiply L by $P(2)$; and so on:

$$\begin{aligned} P(2) &= LP(1) = L(LP(0)) = L^2P(0) \\ P(3) &= LP(2) = L(L^2P(0)) = L^3P(0) \\ &\dots \\ P(t) &= L^tP(0) \end{aligned}$$

Note that the formula for $P(t)$ mirrors the solution $p(t) = r^tp(0)$ of the model $p(t+1) = rp(t)$ in (12.1), with the difference that the powers of the number r have been replaced by powers of the matrix L.

How do we calculate the powers of L that we need to find $P(t)$?

Example 12.3 offers a solution to this problem. It uses powers of *real numbers* instead of the powers of *matrices* to carry out the calculations. To apply the result of the example on the vector $\mathbf{u} = P(0)$, we need to express $P(0)$ as a linear combination of the eigenvectors

$$\mathbf{v}_1 = \begin{bmatrix} 6 \\ 1 \end{bmatrix} \quad \text{and} \quad \mathbf{v}_2 = \begin{bmatrix} -3 \\ 4 \end{bmatrix}$$

that we found in Example 12.2. In other words, we need to find real numbers α_1 and α_2 so that

$$P(0) = \alpha_1 \begin{bmatrix} 6 \\ 1 \end{bmatrix} + \alpha_2 \begin{bmatrix} -3 \\ 4 \end{bmatrix}$$

In Exercise 21 we show that $\alpha_1 = 15$ and $\alpha_2 = 5$; i.e.,

$$P(0) = \begin{bmatrix} 75 \\ 35 \end{bmatrix} = 15 \begin{bmatrix} 6 \\ 1 \end{bmatrix} + 5 \begin{bmatrix} -3 \\ 4 \end{bmatrix}$$

Using (12.9) with $\mathbf{u} = P(0)$, we obtain (recall that $\lambda_1 = 1.2$ and $\lambda_2 = -0.15$)

$$\begin{aligned} P(1) &= LP(0) = \alpha_1\lambda_1\mathbf{v}_1 + \alpha_2\lambda_2\mathbf{v}_2 \\ &= 15(1.2) \begin{bmatrix} 6 \\ 1 \end{bmatrix} + 5(-0.15) \begin{bmatrix} -3 \\ 4 \end{bmatrix} \\ &= \begin{bmatrix} 108 + 2.25 \\ 18 - 3 \end{bmatrix} = \begin{bmatrix} 110.25 \\ 15 \end{bmatrix} \end{aligned}$$

which is what we obtained in (12.7). Using Example 12.3 we can now calculate any power of L, and thus any value of $P(t)$. By using (12.11), we obtain

$$\begin{aligned} P(t) &= L^tP(0) = \alpha_1\lambda_1^t\mathbf{v}_1 + \alpha_2\lambda_2^t\mathbf{v}_2 \\ &= 15(1.2)^t \begin{bmatrix} 6 \\ 1 \end{bmatrix} + 5(-0.15)^t \begin{bmatrix} -3 \\ 4 \end{bmatrix} \end{aligned} \tag{12.12}$$

Note that we could not have obtained (12.12) by merely calculating the powers of L as we started in (12.7). Even if we computed a few more values of $P(t)$ (as we did for $t = 1, 2, 3, 4$), we would not be able to guess a pattern that we could use to find a general formula for $P(t)$.

What can we deduce from (12.12)? As t increases, the exponential term $(1.2)^t$ increases. The absolute value of the second term, $(0.15)^t$, decreases quickly: for instance, $(0.15)^6 \approx 0.000011$ and $(0.15)^{10} \approx 5.7 \cdot 10^{-9}$; this is not surprising, since $(0.15)^t \to 0$ as $t \to \infty$.

We conclude that as t increases, the second term in (12.12) loses its significance, and therefore

$$P(t) \approx 15(1.2)^t \begin{bmatrix} 6 \\ 1 \end{bmatrix} \tag{12.13}$$

Thus, in the long term, the growth of the age groups is determined by the eigenvalue $\lambda_1 = 1.2$: each age group increases approximately by a factor of 1.2 (hence the convergence of the graphs in Figure 12.4c toward 1.2).

The eigenvector

$$\begin{bmatrix} 6 \\ 1 \end{bmatrix}$$

gives the *stable age distribution:* $6/(6+1) \approx 0.857$ is the ratio of minors and $1/(6+1) \approx 0.143$ is the ratio of adults in the total population (see Figure 12.4b).

In conclusion:

(1) It can be proven that a well-defined Leslie matrix of any size has only one positive eigenvalue, and the corresponding eigenvector has positive coordinates. In our case, the eigenvalue is $\lambda = 1.2$, and the corresponding eigenvector is $\mathbf{v} = [6 \quad 1]$. (By "well-defined" we mean that the birth parameters and survival probabilities need to satisfy certain conditions; in the case of a 2×2 Leslie matrix, the survival probability has to be positive (not zero), and at least one age group has to produce offspring.)

(2) The positive eigenvalue λ_1 determines the behaviour of each age group (and hence of the whole population) in the long term. If $\lambda_1 > 1$, then the size of each age group increases, and if $\lambda_1 < 1$, then all age groups shrink in size.

(3) The relative growth rates within each age group approach λ_1 over time.

(4) The eigenvector $\mathbf{v}_1$ corresponding to the positive eigenvalue λ_1 gives the stable age distribution: if $\mathbf{v}_1 = [x \quad y]$, then the ratios of the two age groups in the stable age distribution are $x/(x+y)$ and $y/(x+y)$.

Summary

To study how a certain population develops through its lifespan, we use the **Leslie model.** We define a vector that contains the population sizes of each **age group** (or age class). Based on the **survival probabilities** and the **birth parameters,** we construct the **Leslie matrix.** By applying the Leslie matrix to the vector of age distributions we calculate future age group sizes. The positive eigenvalue of the Leslie matrix determines the growth pattern in the long term, and the corresponding eigenvector gives the relative size of each age group when the population nears its steady state.

12 Exercises

1. The Leslie matrix for a population with three age groups is given by

$$L = \begin{bmatrix} 0 & 2 & 3 \\ 0.5 & 0 & 0 \\ 0 & 0.3 & 0 \end{bmatrix}$$

(a) Explain the meaning of each non-zero entry in L. Explain the meaning of the entry 0 in the first row and the first column of L.

(b) If the initial population is given by $P(0) = [120 \quad 200 \quad 30]$, find $P(1)$ and $P(2)$.

2. The Leslie matrix for a population with three age groups is given by

$$L = \begin{bmatrix} 0.6 & 1.1 & 0 \\ 0.9 & 0 & 0 \\ 0 & 0.1 & 0 \end{bmatrix}$$

   (a) Explain the meaning of each non-zero entry in L. Explain the meaning of the entry 0 in the first row and the third column of L.

   (b) If the initial population is given by $P(0) = [100 \quad 100 \quad 100]$, find $P(1)$ and $P(2)$.

3. What happens to a population whose Leslie matrix is given by the following?

$$L = \begin{bmatrix} 0 & 0 & 0 \\ 0.5 & 0 & 0 \\ 0 & 0.5 & 0 \end{bmatrix}$$

4. What happens to a population whose Leslie matrix is as follows?

$$L = \begin{bmatrix} 2 & 3 & 4 \\ 0 & 0 & 0 \\ 0 & 0.6 & 0 \end{bmatrix}$$

5. Give a reason why the matrix

$$L = \begin{bmatrix} 2 & 3 & 4 \\ 1.1 & 0 & 0 \\ 0 & 0.6 & 0 \end{bmatrix}$$

   cannot be a Leslie matrix.

6. What population dynamics is implied by the following Leslie matrix?

$$L = \begin{bmatrix} 2 & -0.5 & 4 \\ 0.6 & 0 & 0 \\ 0 & 0.3 & 0 \end{bmatrix}$$

7. Consider the Leslie matrix

$$L = \begin{bmatrix} 0 & 0 & 3 & 0 \\ 0.2 & 0 & 0 & 0 \\ 0 & 0.6 & 0 & 0 \\ 0 & 0 & 0.4 & 0 \end{bmatrix}$$

   (a) Explain the dynamics of reproduction implied by L.

   (b) Which age group has the highest mortality?

   (c) Given the initial population $P(0) = [100 \ 200 \ 150 \ 40]$, find $P(1)$ and $P(2)$.

8. Consider the Leslie matrix

$$L = \begin{bmatrix} 2 & 1 & 0 & 0 \\ 0.5 & 0 & 0 & 0 \\ 0 & 0.2 & 0 & 0 \\ 0 & 0 & 0.1 & 0 \end{bmatrix}$$

   (a) Explain the dynamics of reproduction implied by L.

   (b) Which age group has the highest mortality? Lowest mortality?

   (c) Given the initial population $P(0) = [200 \ 60 \ 40 \ 40]$, find $P(1)$ and $P(2)$.

9–12 ▪ For each Leslie matrix,

(a) find both eigenvalues

(b) give a biological interpretation of the positive eigenvalue

(c) find the stable age distribution

9. $A = \begin{bmatrix} 0 & 2 \\ 0.32 & 0 \end{bmatrix}$

10. $A = \begin{bmatrix} 0 & 3 \\ 0.48 & 0 \end{bmatrix}$

11. $A = \begin{bmatrix} 1.2 & 0.5 \\ 0.26 & 0 \end{bmatrix}$

12. $A = \begin{bmatrix} 0.1 & 2 \\ 0.28 & 0 \end{bmatrix}$

13. Consider the Leslie matrix

$$L = \begin{bmatrix} 0 & 1 \\ 0.5 & 0 \end{bmatrix}$$

Describe the dynamics given by L. Using the initial population $P(0) = [100 \quad 100]$, find $P(t)$ for $t = 1, 2, 3, \ldots, 10$. What happens to the population in the long term?

14. Consider the Leslie matrix

$$L = \begin{bmatrix} 0 & b \\ 0.5 & 0 \end{bmatrix}$$

Describe the dynamics given by L. Using the initial population $P(0) = [m \quad n]$, find $P(t)$ for $t = 1, 2, 3, 4$. Based on the pattern you noticed, find a general formula for $P(t)$. What happens to the population in the long term if $b < 2$? If $b = 2$? If $b > 2$?

15. Let

$$L = \begin{bmatrix} 2 & 0 \\ 0 & 3 \end{bmatrix} \quad \text{and} \quad \mathbf{u} = \begin{bmatrix} 5 \\ 6 \end{bmatrix}$$

(a) Find the eigenvalues λ_1 and λ_2 of A and the corresponding eigenvectors $\mathbf{v}_1$ and $\mathbf{v}_2$.

(b) Express the vector $\mathbf{u}$ as a linear combination of $\mathbf{v}_1$ and $\mathbf{v}_2$.

(c) Use your answers to (a) and (b) and formula (12.11) to find $L^8\mathbf{u}$.

16. Let

$$L = \begin{bmatrix} 3 & 1 \\ 0 & 4 \end{bmatrix} \quad \text{and} \quad \mathbf{u} = \begin{bmatrix} -2 \\ -4 \end{bmatrix}$$

(a) Find the eigenvalues λ_1 and λ_2 of A and the corresponding eigenvectors $\mathbf{v}_1$ and $\mathbf{v}_2$.

(b) Express the vector $\mathbf{u}$ as a linear combination of $\mathbf{v}_1$ and $\mathbf{v}_2$.

(c) Use your answers to (a) and (b) and formula (12.11) to find $L^{12}\mathbf{u}$.

17. Let

$$L = \begin{bmatrix} 1 & 2 \\ 3 & 2 \end{bmatrix} \quad \text{and} \quad \mathbf{u} = \begin{bmatrix} 4 \\ 1 \end{bmatrix}$$

(a) Find the eigenvalues λ_1 and λ_2 of A and the corresponding eigenvectors $\mathbf{v}_1$ and $\mathbf{v}_2$.

(b) Express the vector $\mathbf{u}$ as a linear combination of $\mathbf{v}_1$ and $\mathbf{v}_2$.

(c) Use your answers to (a) and (b) and formula (12.11) to find $L^{20}\mathbf{u}$.

18. Let

$$L = \begin{bmatrix} 1 & 0 \\ 3 & 2 \end{bmatrix} \quad \text{and} \quad \mathbf{u} = \begin{bmatrix} 2 \\ -3 \end{bmatrix}$$

(a) Find the eigenvalues λ_1 and λ_2 of A and the corresponding eigenvectors $\mathbf{v}_1$ and $\mathbf{v}_2$.

(b) Express the vector $\mathbf{u}$ as a linear combination of $\mathbf{v}_1$ and $\mathbf{v}_2$.

(c) Use your answers to (a) and (b) and formula (12.11) to find $L^{10}\mathbf{u}$.

19. Let

$$L = \begin{bmatrix} 7 & 0 \\ 0 & -3 \end{bmatrix} \quad \text{and} \quad \mathbf{u} = \begin{bmatrix} 2 \\ 4 \end{bmatrix}$$

Find $L^{13}\mathbf{u}$.

20. Let

$$L = \begin{bmatrix} 5 & 0 \\ 1 & -2 \end{bmatrix} \quad \text{and} \quad \mathbf{u} = \begin{bmatrix} 7 \\ 3 \end{bmatrix}$$

Find $L^9\mathbf{u}$.

Index